Geology of the country around Blackpool

The district described in this memoir lies on the western side of the Fylde and also on the eastern flank of the East Irish Sea Basin. The latter was a major pull-apart basin in Triassic times, in which sedimentation was largely controlled by large north-trending faults. The solid rocks are heavily mantled by glacial deposits and are known only from boreholes. The Triassic deposits include the commercially important Preesall Salt within the Mercia Mudstones; it is preserved in a syncline, one of several folds in the area. The history of the salt mining and brine extraction which led to the establishment of the chemical industry on the west bank of the Wyre is described, and a penecontemporaneous tectonic model is invoked to explain variations in salt thickness. The remaining Triassic sequence includes the Sherwood Sandstone, which, 40 km offshore, is the reservoir rock in the methane gas field in the central East Irish Sea Basin. The hydrocarbon potential of the Blackpool district is discussed. The overlying Mercia Mudstones, containing a wealth of sedimentary structures, are newly correlated and figured.

The glacial deposits of Blackpool North Shore were well exposed in the cliffs before the sea defences were built and the section has been recreated from the century-old notebooks of Charles De Rance. A wealth of new borehole data beneath Blackpool has allowed detailed sections to be drawn across the town, now one of England's foremost coastal resorts and recently the centre of considerable population growth. The sections show, in particular, the sequence in the glacial drifts, the nature of the kettle holes and the shape of the terrain across which the Flandrian marine alluvium was deposited. A detailed account is given of the extensive tract of marine alluvium in the north of the area, with its old storm beaches, now kilometres inland.

Plate 1   Blackpool and the Fylde coast, looking north-east (*Hunting Aerofilms*)

BRITISH GEOLOGICAL SURVEY

A A WILSON and
W B EVANS

CONTRIBUTOR
Palaeontology
G Warrington

Geology of the country around Blackpool

Memoir for 1:50 000 geological sheet 66
New Series (England and Wales)

LONDON: HMSO   1990

First published 1990

ISBN 0 11 884462 8

Bibliographical reference

WILSON, A A, and EVANS, W B. 1990.   Geology of the country around Blackpool.   *Memoir of the British Geological Survey*, Sheet 66 (England and Wales).

Authors

A A Wilson, BSc, PhD
British Geological Survey, Keyworth

W B Evans, BSc
Formerly British Geological Survey

Contributor
G Warrington, BSc, PhD
British Geological Survey, Keyworth

Other publications of the Survey dealing with this district and adjoining districts

BOOKS

Memoirs

British Regional Geology
The Pennines and adjacent areas (3rd edition)

MAPS

1:625 000
Solid geology (south sheet)
Quaternary geology (south sheet)
Aeromagnetic map (south sheet)

1:250 000
Solid geology, Liverpool Bay
Seabed sediments, Liverpool Bay
Bouguer gravity anomaly, Liverpool Bay
Aeromagnetic anomaly, Liverpool Bay

1:50 000
Barrow in Furness (sheet 58) Solid with Drift
Blackpool (sheet 66) Solid with Drift
Garstang (sheet 67) Solid with Drift (in press)
Southport (sheet 74) Solid with Drift
Preston (sheet 75) Solid
Preston (sheet 75) Drift

Printed in the UK for HMSO
Dd 240435   C10   06/90

CONTENTS

FIGURES

PLATES

PREFACE

The first survey at a scale of six inches to one mile of the Blackpool Sheet was made by C E De Rance in 1869 and the map was published in 1871. A short explanatory memoir followed in 1875, and a specialised memoir on the glacial drifts of the coasts of south-west Lancashire, also by De Rance, was published in 1877. In 1921, R L Sherlock described the contemporary knowledge of the Preesall Salt in his report on Rock Salt and Brine.

The resurvey of the sheet in 1968 was carried out by W B Evans and A A Wilson, except for a 1 km-wide strip along the southern edge, surveyed by T H Whitehead and R C B Jones in 1936. The initials of the responsible officers are shown on p.viii against the list of six-inch maps which cover the sheet.

On completion of the mapping, the Geological Survey sank six boreholes between 1969 and 1974 to elucidate the Triassic succession. Two further boreholes were sunk by the Survey in 1982 in connection with a geothermal energy programme. All eight borehole cores were logged by Dr Wilson. The section of the memoir describing the Preesall Salt was written by Mr Evans and those describing the remaining Triassic stratigraphy by Dr Wilson. Dr G Warrington has contributed a section on palynology. The chapters on the extensive drift sequence are by Mr Evans and Dr Wilson, and Mr D J Lowe prepared the figure of the Blackpool cliff section from the De Rance notebooks. Dr G A Kirby advised on the construction of Figure 15.

The account is heavily dependent on borehole records and we acknowledge our indebtedness to Imperial Chemical Industries plc for releasing hitherto confidential borehole data and for providing the photograph for Plate 16. The Borough Engineer of Blackpool Corporation has very kindly given access to a formidable, fully curated database of more than 3000 foundation and sewer boreholes, which has proved invaluable in plotting cross sections and contour maps beneath the town. The extensive borehole records provided by Sub Soil Surveys Ltd and the North Western Road Construction Unit have proved extremely useful. Thanks are also due to Fleetwood Corporation, Geo-Research, Allott and Lomax and the Winter Gardens Company for access to borehole records.

Finally, we wish to thank the many owners and tenants of land who have freely allowed us access to their fields and property to carry out surveys and to drill boreholes; we also wish to thank the Head of the Soil Survey of England and Wales for making copies of their field maps available to us.

The memoir has been edited by Dr A J Wadge.

F G Larminie, OBE
Director

British Geological Survey
Keyworth
Nottingham
NG12 5GG

1st March 1990

LIST OF SIX-INCH MAPS

The following list shows the six-inch maps included wholly or partly within the area of Sheet 66 (Blackpool) of the 1:50 000 Geological Map of England and Wales together with the initials of the surveyors and the dates of survey. Uncoloured dyeline copies of these maps may be purchased from the office of the British Geology Survey at Keyworth, Nottingham. They are available for public reference at that office. The surveyors were W B Evans, R C B Jones, T H Whitehead and A A Wilson. National Grid maps are all within the 100 km square SD.

33 NW	Fleetwood	WBE	1968
33 NE	Preesall	WBE	1968
33 SW	Cleveleys	WBE	1968
33 SE	Hambleton	WBE	1968
34 NW	Blackpool North Shore	AAW	1968
34 NE	Singleton	AAW	1968
34 SW	Blackpool South Shore	AAW, RCBJ	1936 – 1968
34 SE	Great Plumpton	AAW, RCBJ, THW	1934 – 1968

CHAPTER 1

Introduction

Few parts of the country have changed more in the last 150 years than the Blackpool district. For centuries, the Fylde was a land of marshes and peat bogs which were gradually reclaimed piece by piece; access was by a network of narrow, meandering lanes that barely improved the general inaccessibility of the region. As the land was brought under cultivation, it was put down to oats and wheat, and later to vegetables, especially potatoes. All went to supply the growing population of Preston and the other mill towns and cities of Lancashire. There were few visitors to the area, though by 1750 a few stayed in the hamlet of Blackpool, and by the mid-1820s up to three coaches a day made the journey there in the summer along the road from Preston. The most important town, however, was Poulton-le-Fylde at the navigable head of the Wyre, which was the only good route into the heart of the Fylde. In the 18th and early 19th centuries, flax from Ireland and timber from the Baltic came into its anchorages at Skippool and Wardleys but, even so, economic development fell well behind that of the rest of lowland Lancashire.

The awakening, when it came, was due wholly to the coming of the railway, which reached Poulton-le-Fylde and Fleetwood in 1840, and Blackpool in 1846. The coming change was anticipated by Peter Hesketh Fleetwood, who engaged Decimus Burton to lay out the new town of Fleetwood on a waste of sand-dunes. While the town was being built, the railway arrived and an improved harbour and wharves were soon constructed. The original plan was to ferry railway passengers across Morecambe Bay, so avoiding the steep climb through Cumbria that was expected to tax railway engineers and engines to the utmost in constructing and operating the west-coast route from London to Scotland. In the event, this plan was still-born for the line over Shap was completed surprisingly quickly, but ferry services soon began to Belfast and the Isle of Man, and by the time these routes were abandoned, the fishing industry was well established, leading to the eventual construction of extensive fish-processing plants. The industrial nature of the lower Wyre was accentuated in 1872 by the discovery of rock-salt (halite) at Preesall on the eastern bank of the river. Brine was pumped across the estuary and a salt-works was soon established near Fleetwood; an increasingly sophisticated chemical industry has been an important feature of the area ever since.

The development of Blackpool, with no sheltered anchorage and no mineral resources, followed a very different route. Its initial growth was based on the popularity of the new fashion for sea-bathing and, later, on the holiday trade in all its forms. Even before the railway reached Blackpool, scattered buildings extended for over 2 km along the cliff-top, and masonry walls had been built to arrest erosion of the cliffs by the sea. But the railway changed the entire scale of the enterprise and, during the second half of the 19th century, Blackpool became the playground of the mill-workers of

Lancashire and, indeed, of the workers of much of northern England. They came to the town not only for the sea and sands but for the attraction of the Winter Gardens, the Opera House, the electric tramway, the Tower and the illuminations. By the mid-1980s, nearly 6 million passengers visited Blackpool by rail each year. Between the wars, the town expanded mainly to the south of the town-centre; latterly, housing has spread northwards through Cleveleys and now is almost continuous with the southern and western margins of Fleetwood, so that the conurbation (Plate 1) takes in most of that part of the district that lies west of the Wyre and the Skippool Channel (Figure 17). The improvement in transport has recently spread to the roads; with the opening of the M55 link to the M6, it is now possible for cars and coaches to reach Blackpool more quickly and in greater numbers than ever before.

East of the river, the changes have been far less dramatic. The railway did not reach Knott End-on-Sea until 1908, and by 1930 it had closed. Perhaps if it had come sooner, the chemical industry might have grown up on the east bank of the Wyre rather than at Fleetwood. As it is, except for Hambleton which is rapidly becoming a commuter village for the urban area across the Wyre, and Knott End-on-Sea where retirement homes are increasingly popular, this part of the district has remained rural. Dairy farming, poultry-rearing and market gardening are now far more important than cereals. While the two banks of the Wyre are linked only by Shard Bridge, this contrast between an urban west bank and a rural east bank is likely to be maintained.

HISTORY OF RESEARCH

Not surprisingly, the Blackpool district has attracted little geological interest for exposures are poor and most of the deeper borehole results have, until recently, been commercially confidential. The only area of general geological interest seems to have been the Blackpool cliffs.

There is an early record of the occurence of marine shells in the glacial sands of the cliffs (Thornber, 1837), and a detailed account of the section appeared soon afterwards (Binney, 1852). The publication of the first Solid edition of Sheet 66 of the one-inch geological map followed in 1871, and a Drift edition was produced in 1874. Together with the sheet memoir (De Rance, 1875) and a regional evaluation of the drift deposits (De Rance, 1877), these effectively summarised geological knowledge of the district at that date. By then, the drift succession was known in fair detail but drilling to prove the solid rocks had scarcely begun.

The first account of the salt deposits of the district was given by Thompson (1908). A further description was included in one of the Geological Survey's economic memoirs (Sherlock, 1921) though, because the process of natural salt solution was not then understood, this account misrepresents

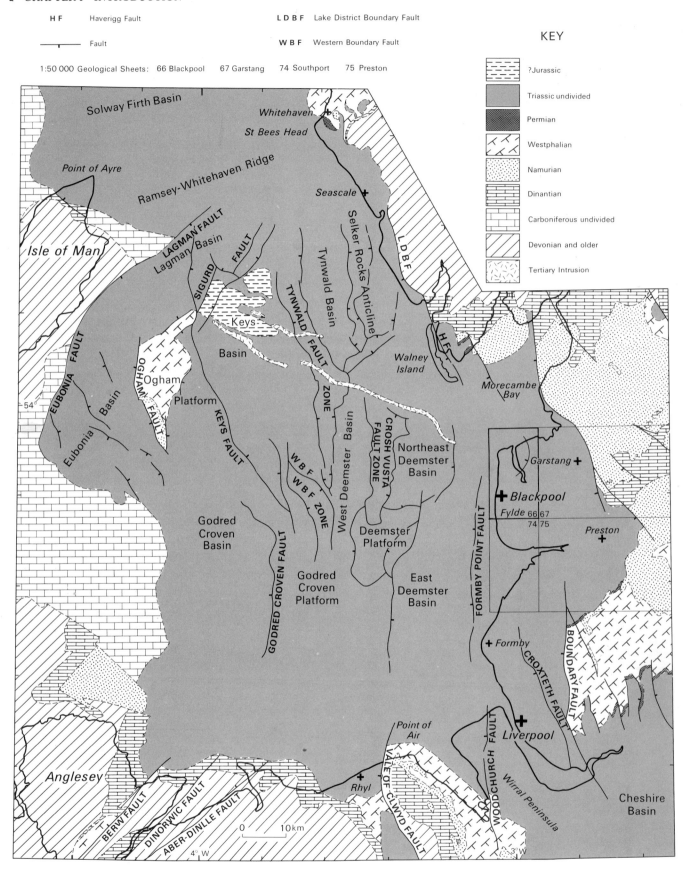

Figure 1   The geology of the Blackpool district in relation to the East Irish Sea Basin (after Jackson et al., 1987)

the geometry of the salt body. No further work was published until Audley-Charles (1970, pl. 2), in a paper on Triassic correlations in Britain, produced a generalised vertical section of one of the boreholes in the saltfield (B6), and implied correlations between the Preesall Salt and the lower of the two main salts of Cheshire, and between a few metres of haselgebirge (halite–mudstone rock) and the upper salt. Even at that date, however, the sequence had not been established with certainty and W B Evans (1970, fig. 8) produced a tentative cross-section that suggested that salts proved west of the Wyre might lie well above the horizon of the Preesall Salt.

To remove these uncertainties, particularly in view of the planned exploration for hydrocarbons beneath the Irish Sea, the Geological Survey resurveyed the Blackpool district in 1968, and supported the mapping with a limited drilling programme. The stratigraphical results of the new holes were published in the annual reports of the Institute of Geological Sciences, (1972, 1975), IGS Boreholes 1974 (1975) and Burgess et al. (1984), incorporated in the 1:50 000 Sheet NS 66, and summarised in the Outline of Geology (Evans and Wilson, 1975) on the published map. The detailed palynological results were published separately by Warrington (1974a).

The only other modern publications dealing with the Blackpool district have been concerned with the Quaternary sequence, and several have touched on individual sites within, or close to, the district, though usually within a broader context. Notable amongst these are papers by Bradbury (1971), Gresswell (1953; 1957; 1967), Oldfield (1956), Oldfield and Stratham (1965), and Tooley (1969; 1971; 1976). The most recent publications are two papers on an elk skeleton found in a kettle-hole near Blackpool (Barnes et al., 1971; Hallam et al., 1973), and a summary of the Quaternary history of lowland Lancashire (Longworth *in* Johnson, 1985).

GEOLOGICAL SEQUENCE

There are no surface exposures of the solid rocks within the district. The range of known strata is small and it is set out in tabular form on the inside front cover. Predictions about the strata that might underlie beds already proved by drilling, and conjectures about rocks that might have been removed by erosion, depend on knowledge of the sequences cropping out in surrounding districts (Figure 1).

Presumably, the Ordovician and Silurian rocks of Cumbria underlie the district, but they must be at considerable depths and are unproven. Their deposition was followed by substantial Caledonian orogenic movements, so that the Carboniferous sequence probably succeeds them unconformably as elsewhere in northern England. Outcrops to the east of the district suggest that Dinantian and Namurian strata, both probably in deep-water turbidite facies, are present; it is possible, though less certain, that some Westphalian strata are also preserved.

A second phase of structural movements occured in late-Carboniferous times and led to local uplift, but the subsequent structural framework does not seem to have been substantially different from that which controlled sedimentation earlier in the Carboniferous. In both cases, many of the main structures appear to be Caledonian faults or folds reactivated by later stresses. Outcrops and boreholes in the adjacent Garstang district make it practically certain that Permian rocks are present at comparatively shallow depths in the east, and they appear to continue beneath an increasingly thick Triassic cover towards the Wyre. The sequence is likely to include breccia-fans and dune-sands at the base, overlain by marine dolomites, anhydrites and sabkha-type mudstones with rock-salt deposits locally. Deposition may well have been controlled by contemporary movements along growth faults and possibly along fold-axes. The conformably succeeding Triassic rocks are the only solid formations yet proved within the district and are described fully in Chapter 2.

The uppermost Triassic rocks and the lowermost Jurassic rocks were certainly once present but it is not known whether local deposition continued into later Jurassic and Cretaceous times, when the structural setting became increasingly affected by movements consequent upon the opening of the Atlantic. The presence of Chalk in Northern Ireland suggests that sedimentation was active in part of the Irish Sea Basin in Cretaceous times. Subsequently, repeated phases of Tertiary uplift removed any Jurassic or Cretaceous rocks that had been laid down. Not until latest Devensian times does the record recommence, when the most recent of the several Quaternary glaciations left behind the tills and sands that, together with the even more recent marine alluvium, make the Blackpool district, at least temporarily, part of the land-mass of the British Isles rather than part of the bed of the Irish Sea.

CHAPTER 2

Triassic rocks

Because the district is entirely covered by drift, direct knowledge of the solid geology is wholly dependent on deep boreholes. The first of these was drilled at the North Euston Hotel, Fleetwood in 1860 and established the presence of Triassic mudstones; another borehole sunk shortly afterwards at Poulton was the first to record rock-salt (halite) and gypsum in these rocks. In 1872, a drilling programme began at Preesall to test for hematite, for this was known to occur on the opposite shore of Morecambe Bay, and pebbles of hematite are common in the drift around Preesall. Instead of hematite, some of the holes struck rock-salt in commercial quantities, and another entered the Triassic sandstone beneath the salt-bearing strata. By 1885, the full sequence of rock-salt had been proved, leading in 1889 to its exploitation; drilling has continued within the saltfield ever since.

Soon after the Second World War, Imperial Chemical Industries plc sank several holes in unexplored ground to the west of the Wyre. Several of these proved comparatively thin beds of rock-salt which clearly did not equate with the much thicker rock-salt at Preesall and lay much nearer the surface than did the latter along the east bank of the estuary. Because there were no recorded rock-salts below the main sequence at Preesall, and because dips were consistently to the west, Evans (1970) suggested that the sequences proved in these western holes lay above the strata proved in the saltfield. This interpretation was, however, speculative and, to clarify the uncertainty, BGS drilled a number of stratigraphical holes as part of the resurvey. These established that additional beds of rock-salt develop extremely rapidly west of the Wyre between the Sherwood Sandstone and the Preesall rock-salt, and that their position near the surface is due to the presence of a sharp, east-facing structure beneath the estuary. This revised correlation has since been confirmed by palynological studies (Warrington, 1974a),

Boreholes featured in this memoir

	National Grid reference	Registered No. in BGS	Salt industry	BGS	Water	See Appendix	Part of borehole in figures 4–13
B1	3390 4479	SD34SW/2	X			X	X
B4	3692 3870	SD33NE/2	X				X
B5	3628 3871	SD33NE/3	X			X	X
B6	3487 4673	SD34NW/12	X			X	X
B8	3225 4529	SD34NW/18	X			X	X
Blackpool Corp. No. 1	3861 3983	SD33NE/1		X			X
Blackpool Corp. No. 3A	3863 4148	SD34SE/3		X			X
Churchtown	3256 4056	SD34SW/1		X		X	X
Coat Walls	3551 4654	SD34NE/130		X		X	X
E1	3468 4746	SD34NW/4	X			X	X
E5	3554 4419	SD34SE/2	X			X	X
E6	3631 4508	SD34NE/30	X			X	
E7	3561 4492	SD34SE/1	X				X
Hackensall Hall	3498 4679	SD34NW/61		X		X	X
Hambleton	3820 4217	SD34SE/5		X		X	X
Mythop	3647 3499	SD33SE/1		X		X	X
North Euston Hotel	3378 4843	SD34NW/1			X		
P1	3593 4780	SD34NE/126	X			X	X
Poulton-le-Fylde No. 8	3530 4009	SD34SE/7	X				
Preesall No. 2 Shaft	4672 3630	SD34NE/66	X				X
Preesall No. 3 Shaft	3405 4662	SD34NE/62	X				X
Preesall No. 8 Borehole	3535 4713	SD34NE/85	X				
Preesall No. 17 Borehole	3674 4763	SD34NE/98	X				
Preesall No. 25 Borehole	3659 4748	SD34NE/86	X				X
Preesall No. 101 Borehole	3560 4519	SD34NE/2	X			X	
Preesall No. 102 Borehole	3616 4620	SD34NE/38	X				X
Staynall	3562 4438	SD34SE/8		X		X	
Thornton Cleveleys	3314 4409	SD34SW/15		X		X	X
Weeton Camp	3888 3603	SD33NE/9		X		X	X
Winter Gardens	3090 3620	SD33NW/1		X	X		

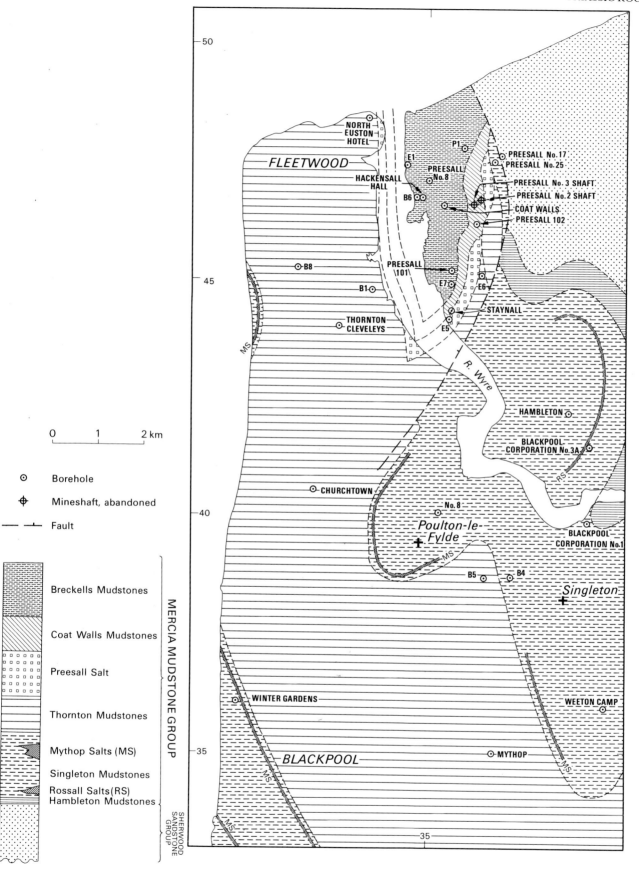

Figure 2 Updated geological map of the Blackpool area showing quoted boreholes

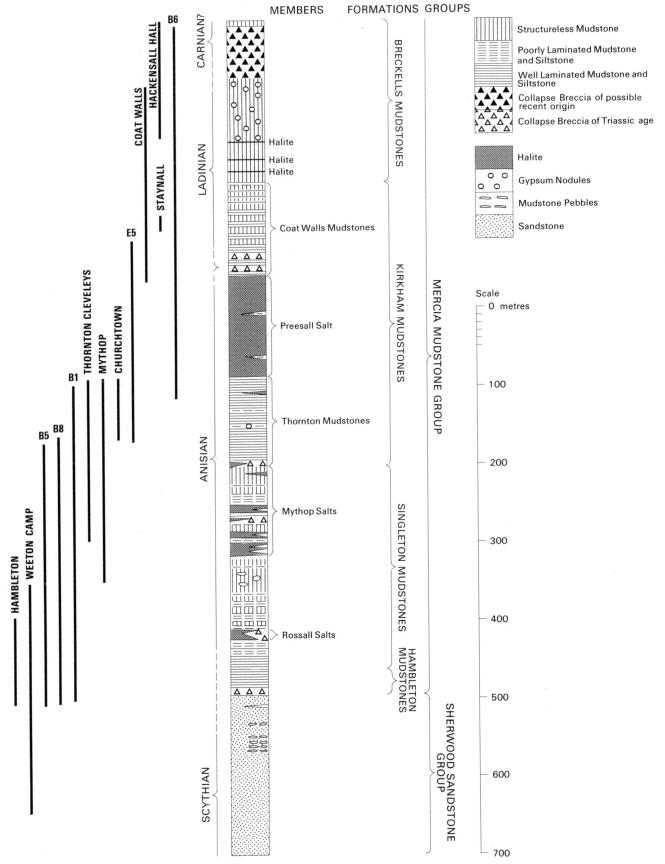

Figure 3   Generalised vertical section of the Triassic strata, with stratigraphical ranges of key boreholes

and further details added as a result of holes drilled for British Gas at Kirkham to the east, for British Gypsum plc at Walney Island to the north, and for BGS near Blackpool, as part of its geothermal exploration programme.

Some of these boreholes are referred to frequently in this chapter and in the accompanying diagrams. Their locations are shown on the geological map that appears on Figure 2, brief logs of them are given in Appendix 1, and their stratigraphical ranges are shown on Figure 3.

CLASSIFICATION

The Triassic sequence is divided into the Sherwood Sandstone and Mercia Mudstone groups, these terms replacing the earlier Bunter Sandstone and Keuper Marl which had chronostratigraphical connotations unsuitable in litho-stratigraphical terms. There is too little information available to subdivide the Sherwood Sandstone Group, though some parts of the sequence are dominantly fluviatile and others are dominantly aeolian. It has, however, proved possible to erect formations within the Mercia Mudstone Group, using criteria similar to those employed by Elliott (1961) around Nottingham. Five formations were initially recognised (Evans and Wilson, 1975), but it is now considered to be more convenient to treat one of these—the Preesall Salt—as a member of the Kirkham Mudstones Formation, and to erect two other members—the Thornton Mudstones and the Coat Walls Mudstones—to cover those parts of the formation that lie respectively below and above the Preesall Salt. The resulting classification is set out in Figure 3.

Though the sequence is almost devoid of macrofossils, palynomorphs have been found in greenish grey bands at many levels in some of the modern boreholes. Five floral assemblages were described by Warrington (1974a) from the first four of the BGS holes. The ages illustrated on the vertical section of the Blackpool sheet (1975) were based upon the assignment of these assemblages to specific Triassic stages. Further work since that date has led to minor revisions and refinements of the boundaries originally proposed (Warrington et al., 1980). The latest positions of the stage boundaries are shown in Figures 3 and 14.

SHERWOOD SANDSTONE GROUP

The Sherwood Sandstone Group crops out beneath thick drift in the north-east of the district and is probably present everywhere else at depth. No single borehole in the Fylde penetrates the entire succession, which is probably at least 300 m thick and may be much more. The local sequence consists almost entirely of fine- to medium-grained and, more rarely, coarse-grained sandstone, dominantly reddish brown but locally grey near the top. Its position in the sequence suggests that it correlates with the Helsby and Wilmslow sandstones of Cheshire.

The thickest proving of the Sandstone was in Preesall No. 17 Borehole, which penetrated 158.47m of reddish brown sandstone with very rare, white and yellow beds; only two mudstone beds were noted, both in the lower part of the se-

0   2   4cm

Plate 2 Cross-bedded sandstone of probable aeolian origin in the Sherwood Sandstone. Weeton Camp Borehole at 249.80 m depth. Individual laminae coloured pale reddish brown or pale grey, show uniform grain-size which differ from layer to layer. (MN 26856)

quence but, since neither the top nor the bottom of the Group was proved, the position of these beds within the overall succession is unknown.

The best modern section is Weeton Camp Borehole which proved the uppermost 141 m of the Group. The sequence consists of three members. The lowest is a fine- and medium-grained sandstone with a few, thin, coarser layers (Figure 4); 'millet-seed' sand grains occur at several levels suggesting a largely aeolian origin (Plate 2), while probable adhesion ripples and cross-bedding are common and may indicate temporary fluviatile conditions. The middle member contains a higher proportion of mudstone, as thin beds and pebbles, and is dominantly of fluviatile origin. The uppermost member consists of fine- to coarse-grained sandstone with common cross-bedding and some probable adhesion ripples. A mudstone bed with desiccation cracks and a band with mudstone pebbles mark water-laid intervals in this dominantly aeolian sandstone sequence. Kirkham Borehole

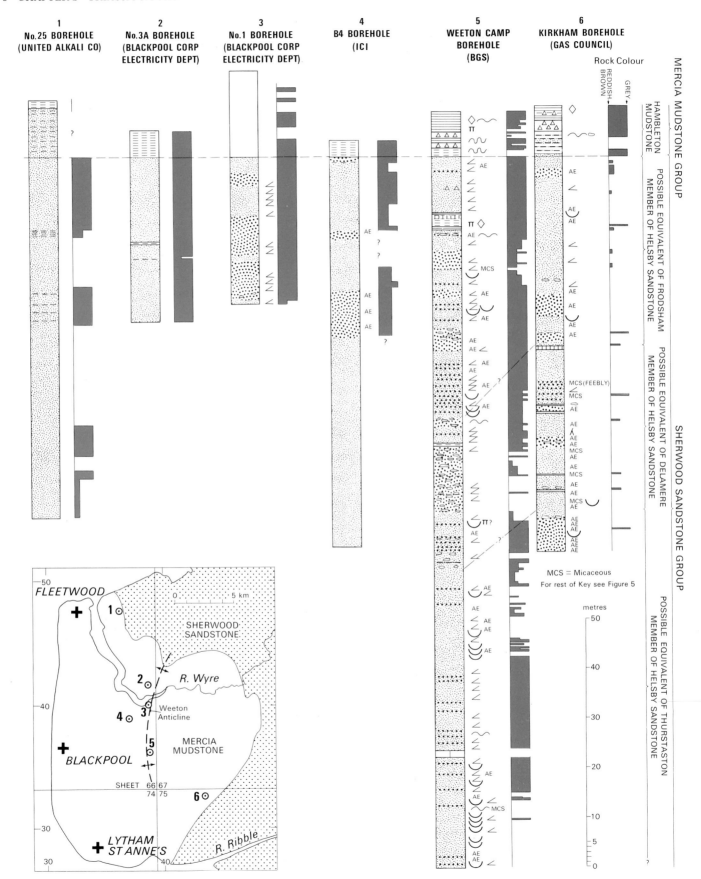

Figure 4 Comparative borehole sections in the Sherwood Sandstone Group

proves a similar sequence (Figure 4). The lowest member is largely medium grained and rich in aeolian grains, but only 9 m were penetrated. The overlying member includes five mudstone beds, but contains fewer mudstone pebbles than the presumed equivalent at Weeton Camp. The highest member has interbedded aeolian and fluviatile sandstones.

These three facies are similar to those of the Helsby Sandstone in north Cheshire (Thompson, 1970b), but there are no reliable markers to prove the detailed correlation. Colter and Barr (in Woodland, 1975) placed much reliance on the higher mudstone content of the middle member, and considered that the three Cheshire facies continue northwards into the Morecambe Bay gasfield.

Other provings have less detailed logs, many of which record only the colour of the sandstone. They show that the topmost part of the Sherwood Sandstone Group varies greatly in colour, probably in part as a consequence of hydrocarbon leaching. In general, boreholes on, or close to, the Weeton Anticline show a preponderance of grey sandstone whilst those around Cleveleys, Preesall Moss and Kirkham are in reddish brown sandstone with rare greenish grey bands.

Details

The thickest sequence of the dominantly grey beds at the top of the Sherwood Sandstone is proved in Weeton Camp Borehole (Figure 4, Section 5),where 126 m of grey sandstone include several reddish brown intervals. The succession is less fully proved in other boreholes, none of which penetrated to the base of the sequence. For example, in B5 Borehole, grey and buff to grey sandstone was cored to 6.8 m and in B4 Borehole (Figure 4, Section 4), grey and pinkish grey sandstone was proved for 35m; a further penetration of 42 m did not specify rock colour. Both boreholes lie on the western flank of the Weeton Anticline. Also on this limb of the fold are Blackpool Corporation Borehole 3A (Figure 4, Section 2), drilled almost exclusively in grey sandstone for 32 m, and Hambleton Borehole, in 6.48 m of greenish grey and grey sandstone banded with hydrocarbon residues. Practically on the crest of the fold is Blackpool Corporation No. 1 Borehole in 30 m of grey sandstone. The basal 0.60 m is brownish grey and may lie in the zone of transition to reddish brown (Figure 4, Section 3).

Boreholes which encountered only reddish brown sandstone at the top of the formation are B1 (to 2.67 m unbottomed), B8 (to 4.62 m unbottomed), and No. 8 Poulton [probably 3529 4009], reputedly drilled in 1837, which encountered 5.38 m (unbottomed) of 'red post' which may also be the topmost Sherwood Sandstone. All the boreholes on Preesall Moss start well down within the Sherwood Sandstone which is almost entirely reddish brown at that level. In Kirkham Borehole (Figure 4, Section 6), the topmost beds were penetrated to a depth of 79 m and were almost exclusively reddish brown.

MERCIA MUDSTONE GROUP

Hambleton Mudstones

The junction between the Sherwood Sandstone and the Mercia Mudstone groups is taken at the incoming of significant mudstone. The contact is in many places associated with an obvious colour change, for the Hambleton Mudstones,the lowest formation of the Mercia Mudstone Group, are dominantly medium grey, contrasting with the reddish

brown or pinky grey Sherwood Sandstone. In some boreholes, however, where the lowermost Hambleton Mudstones locally comprise red mudstones with several thin sandstone bands, there is less contrast in colour.

The main part of the sequence consists of grey mudstone interlaminated throughout with grey siltstones; the latter are micaceous in parts and are 1 cm thick on average. The siltstones generally make up about 10 to 20 per cent of the sequence, and are most common towards the base of the formation. The colour of the sediments is usually a true grey and generally lacks the greenish tinge of the grey beds higher in the Mercia Mudstones. At intervals, there are fine-grained, 'spongy' grey sandstones a few centimetres thick.

0 2 4cm

Plate 3 Giant pseudomorph in calcite after a halite hopper crystal; impression in mudstone in the Hambleton Mudstone. At 111.63 m depth in the Hambleton Borehole. (MLD 8347)

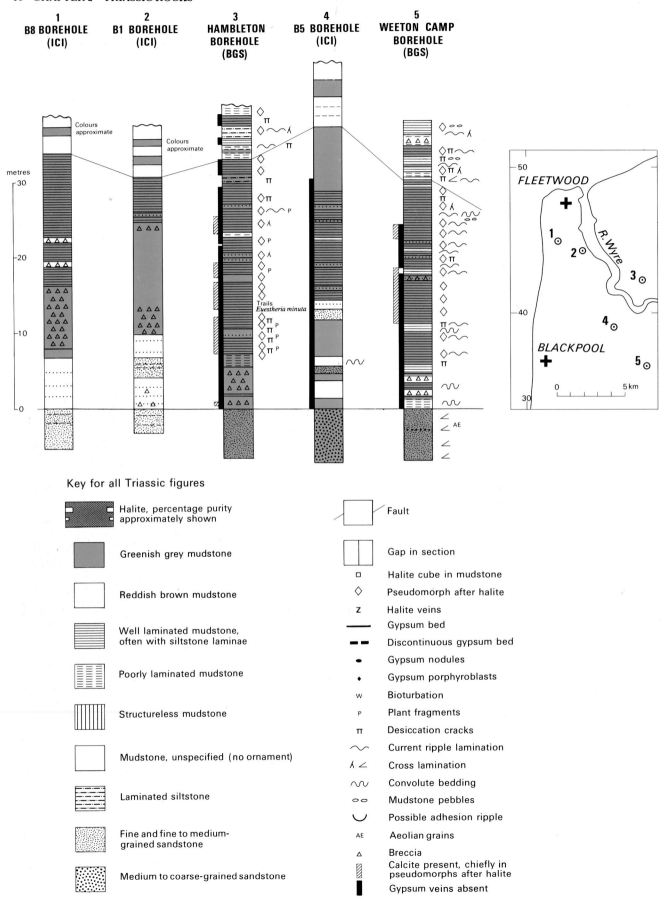

Figure 5   Comparative borehole sections in the Hambleton Mudstones and key to all Triassic figures

These contain voids from which halite crystals have been dissolved, which are now partly filled with calcite. Pseudomorphs after halite, in which the characteristic cubic shape of the salt crystals is preserved in mudstone, are also common in the mudstones, and include many giant specimens more than 10 cm across (Plate 3). These crystals, too, are now replaced by calcite, which also occurs in a few veins and is typical of this part of the sequence; in contrast, gypsum is absent except in veins at the top of the formation (Figure 5). Sedimentary structures are generally well developed and commonly include mudcracks, current-ripples, cross-laminations and clay-flake conglomerates;

current-ripples and flasers were particularly well seen in the cores from Weeton Camp and Kirkham boreholes (Plate 4). In Hambleton Borehole, finely comminuted plant debris occurs at several levels; *Euestheria minuta* and organic trails are also recorded (Figure 5).

Heavy brecciation is common close to the bottom of the grey beds. In the logs of older boreholes, these occurences were recorded as fault breccias (e.g. B1 and B8 boreholes) but recent provings in Hambleton, Kirkham and Weeton Camp boreholes all lie at about this horizon, suggesting that the brecciation is sedimentary in origin (Figure 5). The clasts in the breccia are of grey mudstones interlaminated with

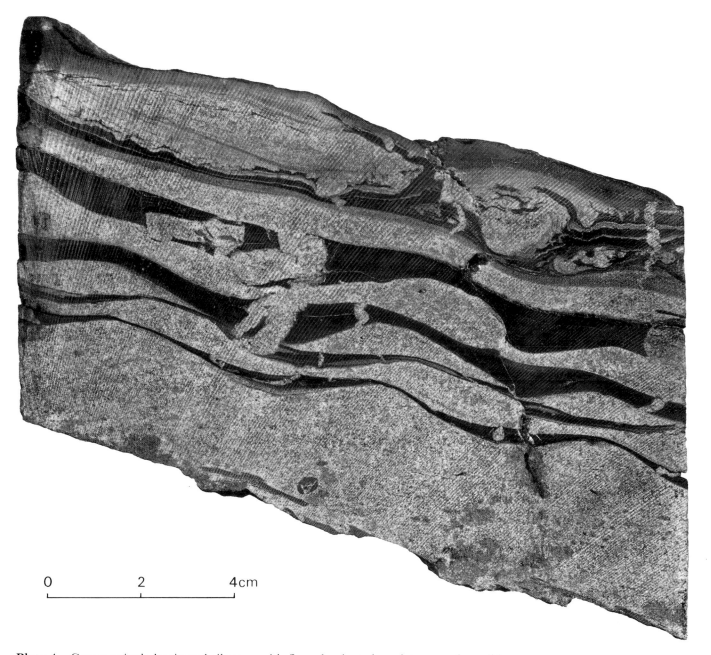

0   2   4cm

Plate 4   Current ripple-laminated siltstone with flaser-laminated mudstones and possible gas-escape structures, in the Hambleton Mudstones. At 348.32–348.40 m depth in the Kirkham Borehole [4324 3247]. (MN 26857)

grey siltstones; they are 10 cm or more long and show plastic deformation and injection of silt (Plate 5). It may be that this widespread brecciation represents thixotropic disruption of soft sediment by an earthquake shock.

Besides those already quoted, several other boreholes have passed through the Hambleton Mudstones. Blackpool Corporation Borehole 3A proved 30.80 m of beds which were wholly grey except for three reddish brown bands each 0.60 m thick. Borehole 8 at Poulton also presumably penetrated the Hambleton Mudstones, but the grey colour is not recorded in the borehole log; it is not recorded either in Borehole 25 at Preesall, though there the formation may be cut out by a fault.

```
0       2       4cm
L       L       L
```

Plate 5 Clasts of interlaminated mudstone and siltstone in a siltstone matrix in the Hambleton Mudstones. There has been thixotropic disruption of sediment, probably due to an earthquake shock. At 131 m depth in the Hambleton Borehole [MLD 8348]

Singleton Mudstones

The succession is dominantly reddish brown mudstone with impersistent beds of halite. The mudstone is commonly structureless or only poorly bedded, but with some better laminated portions interbedded with siltstones. A few beds of greenish grey mudstone occur, and there are also sporadic, greenish grey blotches in parts of the reddish brown mudstone. Gypsum veins occur throughout.

In the present district, the Singleton Mudstones thin eastwards away from the depositional centre of Triassic sedimentation in the main East Irish Sea Basin. From a maximum of 312 m in B1 Borehole, the succession thins inland to 182 m, and to only 104 m in the Kirkham Borehole, near the basin margin.

Beds of rock-salt within the Singleton Mudstones are confined to the west and centre of the Fylde (Figure 6). In the east, in the Kirkham Borehole, no salt was recorded and this area appears to lie outside the original area of salt deposition. Nor was there any salt at Preesall, apart from one possible proving in faulted strata. Farther west, salt beds develop at two levels forming the Rossall Salts near the base of the sequence, and the Mythop Salts near the top.

The Rossall Salts have been found only in B8 Borehole, but their absence from Hambleton Borehole may be due to recent solution, for surface depressions [3776 4364, 3869 4186] closely resembling salt subsidences occur near this borehole, though they are difficult to distinguish geomorphologically from the glacial drainage channels that also occur hereabouts.

The Mythop Salts are thickest in B8 Borehole, but are split into thinner salts, commonly of haselgebirge lithology, in Thornton Cleveleys and Mythop boreholes. The halite recorded in Winter Gardens Borehole is thought to be a bed high in the Mythop Salts (Appendix 1).

Details

About 20 m of more or less well-laminated mudstones with siltstones lie between the top of the Hambleton Mudstones and the horizon of the Rossall Salts and its associated breccias. They are mostly reddish brown but include several beds of greenish grey mudstone with some gypsum nodules in their upper parts. Current-ripple lamination, cross-lamination, desiccation cracks and pseudomorphs after halite are common in these basal beds in the Hambleton, Weeton Camp (Figure 6) and Kirkham boreholes, and they contain *Euestheria minuta* at Kirkham.

Close to the coast near Rossall School, B8 Borehole proved the Rossall Salts towards the base of the formation. Just below the salts are mudstones that contain breccias of probable Triassic age; the breccias may pass laterally into further salts lying nearer the depocentre of the East Irish Sea Basin. The saliferous beds in B8 are 11.50 m thick but they are represented only by breccias in B1 Borehole (Figure 6). Similarly in Hambleton Borehole, a breccia containing gypsum porphyroblasts occurs at this horizon. If the possible subsidence features near this borehole did originate in this way, the halite may only recently have been removed by solution at wet-rockhead, and relict masses may well remain close to the subsidence hollows. Borehole E6 recorded 'grey sand and salt' from 373.99 m to the base of the hole at 384.96 m. It is possible that this record represents the Rossall Salts faulted against the Sherwood Sandstone, with a resultant admixture of halite and sand, but such a reading is highly speculative. Farther south in B4 and B5 boreholes, halite was absent at the horizon of the Rossall Salts.

That part of the Singleton Mudstones lying between the Rossall and Mythop salts has been fully proved in boreholes B1, B5 and B8 (Figure 6). It includes much reddish brown, structureless mudstone with gypsum nodules, passing upwards and downwards into better laminated, reddish brown mudstones with some greenish grey bands. The recent borings at Hambleton, Weeton Camp and Mythop collectively show that the beds overlying a breccia equivalent to the Rossall Salts are more or less well-laminated mudstones with current-ripple lamination, cross-lamination and desiccation cracks, but include some beds of structureless mudstone; the overlying beds grade up by alternation into mainly structureless mudstones with gypsum nodules; these are, in turn, overlain by poorly laminated, largely reddish brown mudstones with some desiccation cracks in their better-laminated portions.

The Mythop Salts are named from Mythop Borehole, where they occur within 59 m of commonly structureless, red mudstone containing halite veins up to 10 cm thick. At intervals, there are beds of haselgebirge (halite-mudstone rock) within the sequence (Plate 6), and at the top of the succession there are almost pure halite beds up to 2 m thick. In those clastic parts of the sequence where bedding is apparent, there are ripple marks, cross-lamination, convoluted lamination, pseudomorphs after halite and rare desiccation cracks; gypsum nodules are very rare. Breccias above and below the Mythop Salts are the likely correlatives of further salt beds preserved nearer the centre of the East Irish Sea Basin.

Towards the coast, the amount of salt in this part of the succession increases. The sequence in Thornton Cleveleys Borehole is at least 101 m thick, with 14 separate beds of salt, each up to 4 m thick, separated by dominantly reddish brown mudstone (Figure 6). As at Mythop, the salt beds all contain intermixed mudstone, commonly as a haselgebirge lithology, but they are generally much thicker than their attenuated equivalents at Mythop. The beds between the salts lithologically resemble the upper member of the Kirkham Mudstones in their close alternation of structureless and poorly laminated mudstone. Desiccation cracks commonly occur, and current-ripples and cross-laminations are also present. In B1 Borehole, only the highest beds of rock-salt seem to be present, and they are largely marly; the lower halite layers of the Mythop Salts are apparently represented by a breccia (Figure 6). Farther west, in B8 Borehole, the Mythop Salts are richer in halite than at any other locality so far drilled. The halite beds tend to be concentrated towards the top and bottom of the saliferous sequence, with a dominantly marly sequence in the middle. They seem to contain less mudstone than in B1 Borehole, and the thickest bed of salt (18.14 m) is fairly pure except for scattered bands of grey mudstone. It is this halite that appears to pass into the breccia in B1 Borehole within a distance of 1.7 km.

The E6, B5, and Thistleton Bridge [4079 3835] boreholes contain little or no Mythop Salt and show that these beds fail eastwards (map in Figure 6). None of them contains more than 2 m of halite in total, though the last borehole terminated within the saliferous sequence, so there may have been further rock-salt at greater depths.

A borehole at Wallpool Bridge [4117 4042] encountered collapsed beds between 52 and 62 m close to the base of the drift; these beds may have been produced by the complete wet-rockhead solution of thin Mythop Salts. In Kirkham Borehole, the equivalent beds are red mudstones varying from structureless to laminated. They are free from halite but include some breccias, the best developed lying between 292.25 and 296.57 m depth and containing fragments up to 8 cm long. It is uncertain whether this breccia represents part of the Rossall, or of the Mythop, salts.

The uppermost part of the Singleton Mudstones, lying above the main group of Mythop Salts, is thickest in the coastal area. The beds are 50 m thick in B1 Borehole where they are dominantly reddish brown, partly banded mudstones with greenish grey bands in their middle and upper parts. Gypsum veins occur at many levels, but halite veins are present only near the base. In B8 Borehole brec-

0   2   4cm

Plate 6 Halite interbedded with siltstone containing halite crystals in the Mythop Salts. At 301.03–301.35 m depth in the Thornton Cleveleys Borehole. (MN 26863)

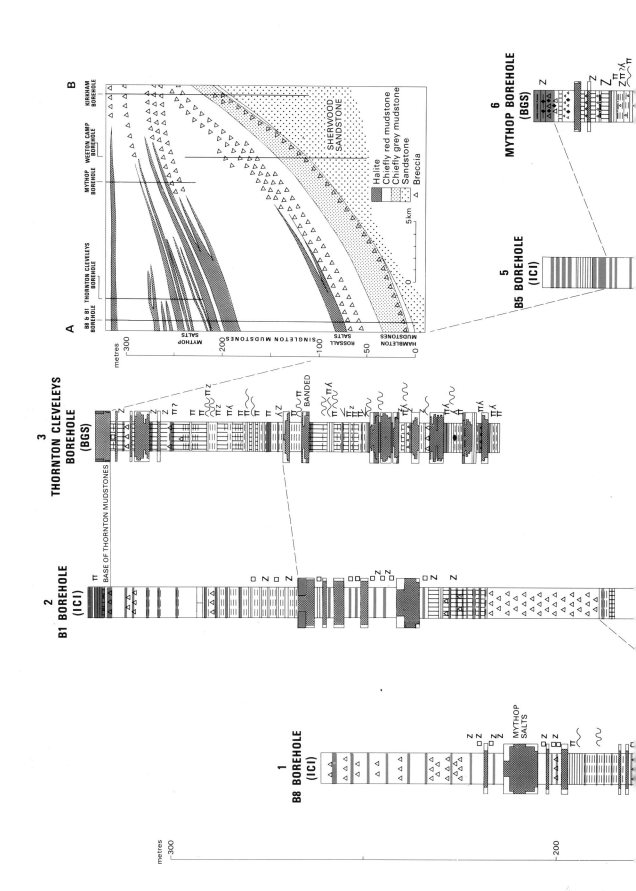

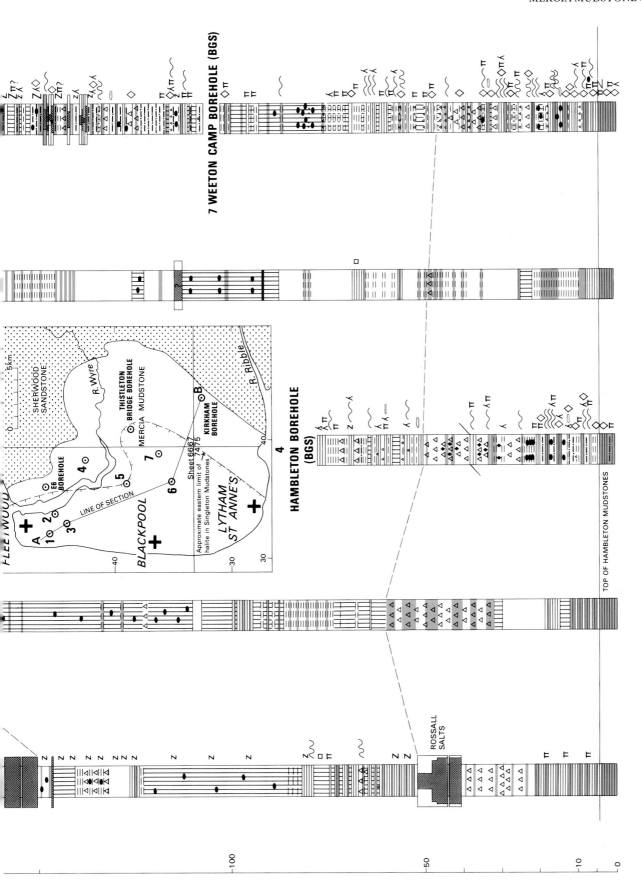

Figure 6 Comparative borehole sections in the Singleton Mudstones, together with borehole site plan and schematic section in the Hambleton and Singleton mudstones, showing the lateral passage of halite beds into breccias of Triassic age towards the basin margin.

ciation is particularly strong, occurring at several horizons (Figure 6). It is likely that these breccias are the equivalents of further salt beds preserved off-shore. In Thornton Cleveleys Borehole four beds of salt with some admixed mudstone in haselgebirge texture occur close to the top of the Singleton Mudstones. These beds totalling 5.18 m in thickness are equivalent to the highest Mythop Salts of Mythop Borehole. Strata between these highest Mythop salts and the main group of salts include both structureless and laminated mudstones; ripple marks and desiccation cracks occur commonly in the latter. These strata vary in thickness from 27 m in Mythop Borehole to 32 m in Thornton Cleveleys Borehole.

Kirkham Mudstones

The Kirkham Mudstones, defined in the Kirkham Borehole section, are characterised by reddish brown and greenish grey mudstones interlaminated with thin siltstones. They commonly show vivid colour-banding, in contrast to the dominantly reddish brown Singleton and Breckells mudstones that respectively underlie and overlie them. In the present district, the Kirkham Mudstones consist of three members. The lowest is the Thornton Mudstones, newly named from the Thornton Cleveleys Borehole section where it consists of well laminated mudstones and siltstones, alternating in reddish brown and greenish grey sections 1 m to 12 m thick. The middle member is the Preesall Salt, which consists almost wholly of rock-salt with only a few thin mudstone bands. Between Preesall and Staynall, extensive collapse-breccias mark the wet-rockhead crop of the Preesall Salt. The highest member, newly named the Coat Walls Mudstones from the BGS borehole section, consists of many alternations of laminated strata and structureless mudstones; it is preserved only in the Preesall and Kirkham synclines.

Thornton Mudstones

The member has recently been proved in BGS boreholes at Churchtown, Mythop and Thornton Cleveleys, as well as in the Gas Council Borehole at Kirkham. In addition, partial sections occur in boreholes B1, B5, B6 and E5 where the logs are less detailed.

Within the present district the Thornton Mudstones are, on average, about 113 m thick, but in Kirkham Borehole they only attain 71 m. Outside the Preesall saltfield they consist of eleven colour-paired cycles, each composed of greenish grey mudstone below, and reddish brown mudstone above, in roughly equal amounts (Figure 7). Usually, numerous siltstone layers about 1 cm thick give a strong banding to the beds. The siltstones are commonly paler tones of the same colour as the adjacent mudstones, which accentuates the banded appearance of the rocks. The eleven major cycles, identified below as Cycles A to K, differ from each other mainly in their thickness. They can be correlated between Thornton Cleveleys, Churchtown, Mythop and Kirkham boreholes and, with less confidence, can be recognised in the less detailed logs of E5 (Figure 7) and B1 boreholes.

Cycle A is locally one of the thickest of the cycles and has been recognised in Thornton Cleveleys, Mythop and B1 boreholes. In Kirkham Borehole, however, it appears to be absent, and the equivalent beds may form the uppermost part of the Singleton Mudstones, making the base of the Kirkham Mudstones locally diachronous. The grey beds of

0   2   4cm

Plate 7 Interlaminated reddish brown siltstones and mudstones with desiccation crack, in the Thornton Mudstones.

Note how the higher sediments curl up near the crack. Cycle H in Thornton Mudstones. Churchtown Borehole at 108.10–108.31 m depth. (MLD 6501)

the cycle include two beds of rock-salt in Thornton Cleveleys Borehole, which may correlate with solution-breccias at this horizon in Mythop Borehole. The overlying red mudstones in Cycle A are thicker than in the other cycles, and contain greenish grey mudstones as elements of minor cycles. Cycles B to G correlate fairly well from borehole to borehole, except at Kirkham where it is possible that the grey beds of Cycles F and G have amalgamated into a single grey sequence. Cycle H is particularly distinctive, being characterised by a major, basal, grey sequence which is split by minor red beds, and overlain by dominantly reddish brown beds including two persistent grey beds, each with subsidiary reddish brown interbeds (Figure 7). Cycles I to K show considerable uniformity throughout the district from Blackpool to Kirkham, though there is local variation in the number of interbeds of a contrasting colour. The red beds of Cycle K are thicker than in any other cycles except A, and also include a higher proportion of structureless mudstone. They are overlain in E5 Borehole by the Preesall Salt, and in the other boreholes of Figure 7 by the collapse-breccia caused by its total solution. Thin beds of rock-salt, commonly of haselgebirge lithology, occur locally in Cycle K, as for instance in B6 and E5 boreholes.

The individual siltstones in the Thornton Mudstones are commonly 2 to 40 mm thick. X-ray analyses show that they are usually dolomitic, and mica occurs rarely on some bedding planes. A visual comparison of the relative proportions of mudstone and siltstone in borehole cores shows that the amount of siltstone decreases towards the depositional centre of the East Irish Sea Basin. Thus, the proportion of siltstone in the sequence is about 50 per cent at Kirkham, 30 per cent at Mythop and only 20 per cent at Thornton Cleveleys and Churchtown.

Sedimentary structures are common (Figure 7). The most characteristic are desiccation cracks, clearly displayed in these well laminated strata (Plate 7). Current-ripple lamination occurs at many levels, particularly in Cycle I at Churchtown, Mythop and Kirkham. Cross-lamination, cut-and-fill structures, and thin convoluted units are common. Gypsum nodules occur at several horizons, and enterolithic folding occurs in places around groups of nodules or plume-shaped masses of gypsum (Plate 8). Veins of gypsum, usually subparallel to the bedding, are generally present, and halite veins are recorded from some boreholes.

Breccias occur sporadically, and there are a few thin beds of salt, notably in Cycle A, and probably in Cycle K in E5 Borehole (Figure 7). Pseudomorphs after halite occur in mudstone at intervals throughout the sequence (Plate 9).

Macrofossils are decidedly rare in the borehole cores. Trace fossils, both trails and borings, were seen at three separate horizons at Mythop, more doubtfully at three others in Thornton Cleveleys, and at one horizon in Kirkham. At a depth of 122.81m in Churchtown Borehole, *Euestheria minuta* occurs abundantly on a single bedding plane in Cycle F (Plate 10). A single plant fragment was recorded from Kirkham.

0   2   4cm

Plate 8   Gypsum near the base of Cycle I in the Thornton Mudstone. Enterolithic folding has caused updoming of reddish brown, interlaminated mudstone and siltstones. At 103.25–103.45 m depth in the Churchtown Borehole. (MLD 6487)

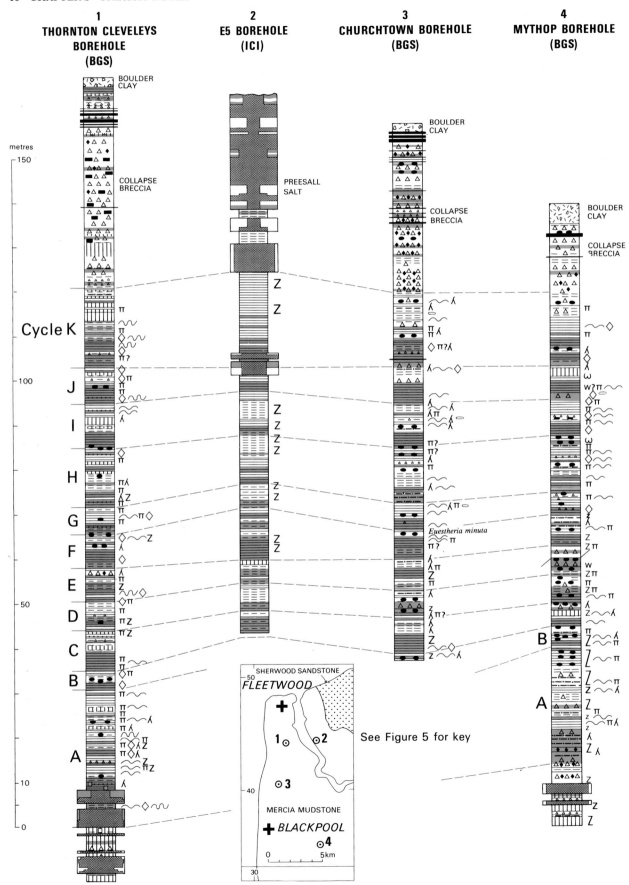

Figure 7 Borehole sections in the Thornton Mudstones

Plate 10   *Euestheria minuta* from Cycle F in the Thornton Mudstones. At 122.81 m depth in the Churchtown Borehole. (MLD 8524)

0   1cm

Plate 9   Small pseudomorphs after halite in greenish grey siltstone from Cycle K in the Thornton Mudstones. At 82.14 m depth in the Churchtown Borehole. Magnification ×3. (MLD 6524)

PREESALL SALT

The thin beds of salt at the top of the Thornton Mudstones are succeeded upwards within a few metres by the Preesall Salt, a major rock-salt member whose presence in the Preesall area has been known for over a century. Its base is taken at the base of the lowest thick salt (Bed A of p.22); where the thin underlying salts are not recorded, this definition is difficult to apply, for the thin bands may have united with the main basal salt, but any consequent errors in correlation are only minor. The salt has been explored by, and exploited from, almost a hundred boreholes within a tract close to the west and south-west of Preesall. The saltfield is currently in the ownership of ICI plc and, although the detailed logs of the boreholes largely remain commercially confidential, the Company has kindly allowed the general findings of the boreholes to be summarised and has also released several detailed borehole logs from their confidential cover.

In depicting the 'outcrop' of the Preesall Salt on the 1:50 000 sheet, the same convention has been adopted as was employed in the Cheshire saltfield (Evans et al., 1968). The complete sequence is preserved only at depth in the dry-rockhead area. As the salt beds rise eastwards towards the surface, they are progressively dissolved by groundwater and never crop out. There is, thus, a belt where only part of the sequence is preserved; this is known as the wet-rockhead area, and its limits, as far as can be determined, are shown as the 'outcrop' of the Preesall Salt despite the fact that the latter usually terminates upwards about 50 to 75 m below the base of the drift, its place being largely taken by brecciated Coats Walls Mudstones lowered by collapse from their true stratigraphical position. The relationship is illustrated by the cross-sections in Figure 8. In practice, the boundaries of the wet-rockhead cannot be plotted with precision, partly because boreholes are less common in this tract than in the main body of the saltfield, partly because original variations in the thickness of the highest salt bed make it locally uncertain whether any solution has taken place, and partly

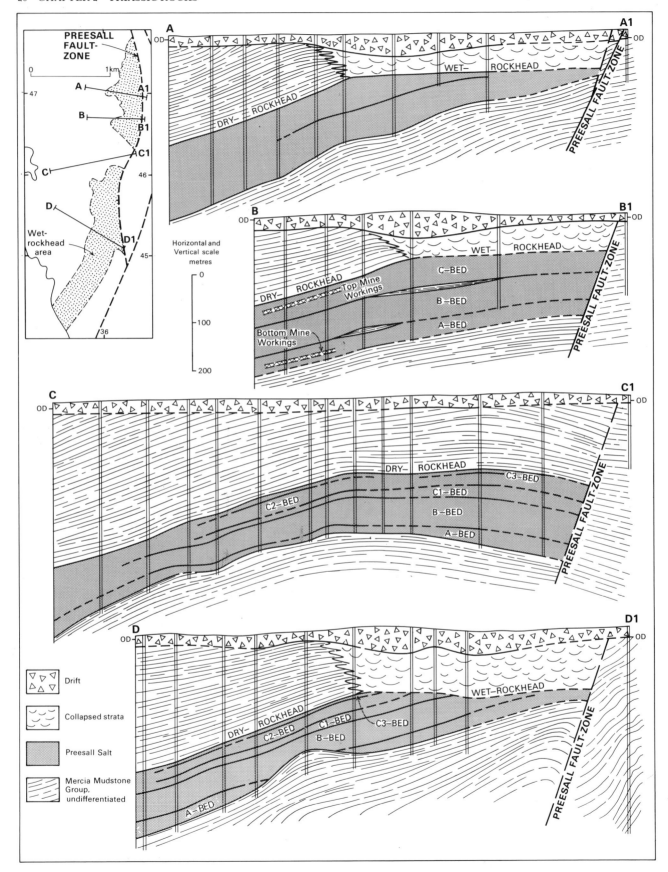

Figure 8 Generalised horizontal sections of the Preesall Salt

because local variations in the transmissibility of the cover-rocks, together with minor flexures and fractures, probably make the boundary highly dentate.

Provings of the sequence are concentrated around the former Preesall salt-mine [about 362 467] and in the working brine-field [358 454] farther to the south-west. The boreholes are sufficiently close to allow structure contours (Figure 9)

and isopachytes of the Preesall Salt to be constructed (Figure 10). Throughout the area near the mine, the salt is about 100 to 130 m thick, with a belt of thinner salt trending north–south, just to the west of the disused mine. To the east, the salt thickens to about 150 m near the Preesall Fault; this is particularly obvious around Park Cottage where the provings are probably closest to the fault-zone. There is a

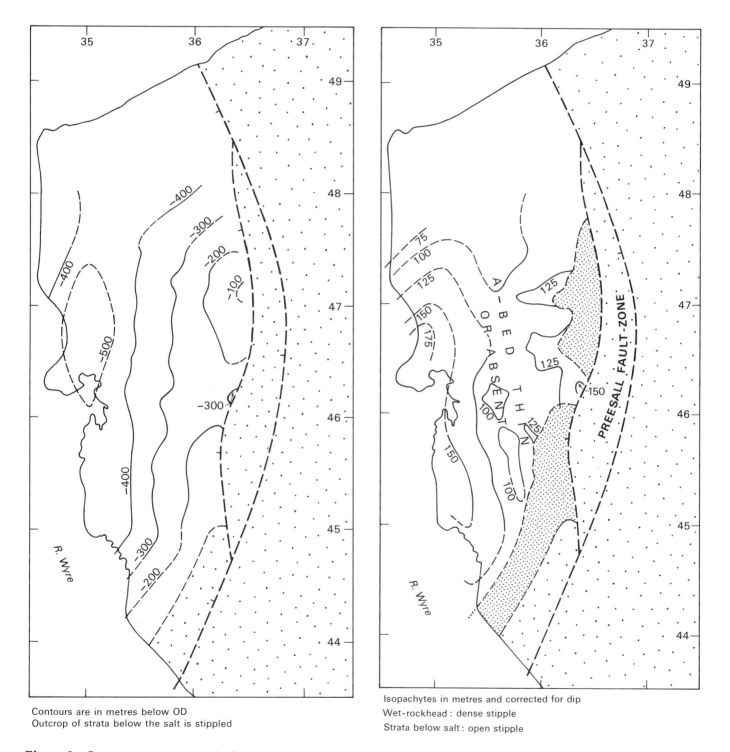

Contours are in metres below OD
Outcrop of strata below the salt is stippled

Isopachytes in metres and corrected for dip
Wet-rockhead : dense stipple
Strata below salt : open stipple

Figure 9 Structure contours on the base of the Preesall Salt

Figure 10 Isopachytes of the Preesall Salt

similar thickening of the salt westwards towards the Wyre, where there are several provings of over 150 m. Indeed, in B6 Borehole almost 180 m of salt are recorded; unfortunately there are no regular records of dips in this borehole to allow this apparent thickness to be corrected to true thickness, but the thin salt beds at the top of the Thornton Mudstones seem to be much their usual thickness, so any correction needed is probably small. The only other proving of the Preesall Salt in the west is in E1 borehole. In contrast to the sequence in B6, this proves an abnormally thin sequence of 81 m, again uncorrected for dip.

The boreholes across the saltfield also provide detailed information on the bed-structure of the Preesall Salt, and illuminate the variations in its thickness. The more recent boreholes establish that the sequence of individual beds within the Preesall Salt is similar throughout the main saltfield, though few of the older holes are recorded in sufficient detail to demonstrate this, particularly since some terminated before reaching the base of the member. Selected vertical sections are given in Figure 11, which also shows the alpha-numerical terminology used in this account.

The sequence is seen at its best in B6 Borehole. Above the thin salt beds at the top of the Thornton Mudstone, three main beds of salt, A-bed, B-bed and C-bed, were proved. They are separated by thin partings of red and grey mudstone; brecciated mudstones are recorded only at the top of A-bed and B-bed where they signify pauses in deposition, possibly with emergence. Subsidiary partings usually split C-bed into three, here called C1-bed, C2-bed and C3-bed, though in B6 Borehole the upper parting cannot be identified with certainty. The general persistence of these partings is remarkable considering that, out of the 183 m of Preesall Salt recorded in B6, the partings together account for only slightly over 7 m, the thickest being 2.1 m between B-bed and C-bed.

The same sequence of beds is clearly seen in the working brine-field, for example in Preesall No 101 Borehole. Northwards into the salt-mine area, the borehole logs are not recorded in sufficient detail to establish the presence of all the partings in any one borehole. It is difficult to separate in the logs the thin, discrete, mudstone partings from beds of marly salt. Even so, there are sufficient records to establish that the individual beds of salt generally persist, and it is significant that the more modern holes are noticeably easier to correlate; for example all the main and subsidiary partings are easily identified in P1 Borehole in the extreme north at Parrox Hall. The previously published records of the sequence thereabouts are the logs of No. 2 and No. 3 shafts at Preesall mine, and these are particularly unsatisfactory. The former probably terminated in the parting between A-bed and B-bed; the latter shows the two main partings, but that between A-bed and B-bed is abnormally thick at 17.5 m. This thickness is not repeated in the other provings around the shaft which, however, demonstrate that the two working levels in the mine were within A-bed and C2-bed, presumably concentrating on the purest bands within these beds (Figure 8).

By plotting the main partings on cross-sections, it is possible to demonstrate that the Preesall Salt thins in the belt trending north–south through the centre of the saltfield mainly because A-bed thins appreciably and even, in places, fails

altogether (Figure 8). The other beds also vary in thickness across this belt, though generally less dramatically. In particular, C-bed, and probably mainly C2-bed, thickens to both the east and the west of the median belt.

Surprisingly, the sediments in the logs are not described in detail. Most of the older records were compiled by drillers, some giving only the depths to the top and bottom of the Preesall Salt. Other, more modern, boreholes have been electrically logged; these logs show the position of the mudstone partings with precision but give no information on the nature of the salt or the partings. Sherlock (1921) quotes an unpublished note by De Rance referring to the rock-salt in the mine as being 'remarkable for [the] occurrence of long sweeps resembling the planes of current-bedding observable in ordinary mechanically formed deposits'. It is possible that he is describing the curved bedding associated with large-scale polygons in the salt (Tucker, 1981), but one cannot be certain. De Rance also mentions that the marls contain 'ripple-marked flags' and 'salt-pseudomorphs', but he does not relate these to any specific horizon; they may refer to the Coat Walls Mudstones. Otherwise there is little to add. Much of the rock-salt is described as 'slightly marly', though other beds were noted as being of high quality and were presumably free of marl. It is even uncertain whether the bulk of the sequence is regularly banded, as much of the salt is in Cheshire (Earp and Taylor, 1986), or whether haselgebirge lithologies are commoner, while De Rance's record of apparent current-bedding presents a third possibility.

There are few boreholes around the western and southern margins of the saltfield. Where the salt thins westwards in Borehole E1, however, the sequence of individual beds of salt becomes unrecognisable, although the hole is only about 850 m from B6. In the south, the only relevant proving is Borehole E5. Again the sequence is difficult to interpret, only partly because the borehole is in the wet-rockhead area. It is not even certain where the base of the Preesall Salt should be taken. Depending on this, the preserved thickness is either 75 m or 95 m (uncorrected for 15° dip), to which must be added the thickness of C-bed and, possibly, much of B-bed. The most notable feature of the log, however, is that a third of the sequence is logged as 'marl and salt'. It seems likely that this represents a real deterioration in the quality of the rock-salt rather than a vagary of the logging.

It is obvious from Figures 9 and 10 that the isopachytes of the Preesall Salt and structure-contours drawn on its base are closely related, even taking into account the imprecise data around the margins of the salt-field. In the east, the thickening towards the Preesall Fault is in precisely the same area as that in which the generally westward dip reverses. The conclusion that downward movement was particularly rapid on the downthrow side of an active fault seems inescapable. Similarly, the considerable thickening westwards into the presumed syncline, and the even more rapid westward thinning of the salt once this axis is crossed, is likely to be due to rapid contemporaneous sagging along the syncline; even if the western limb of the syncline is in reality a fault-zone, the argument for contemporaneous movements is unaffected. It follows that deposition of the Preesall Salt is likely to have been largely confined to a fault-bounded trough, with the greatest thicknesses being laid down close to the margins of

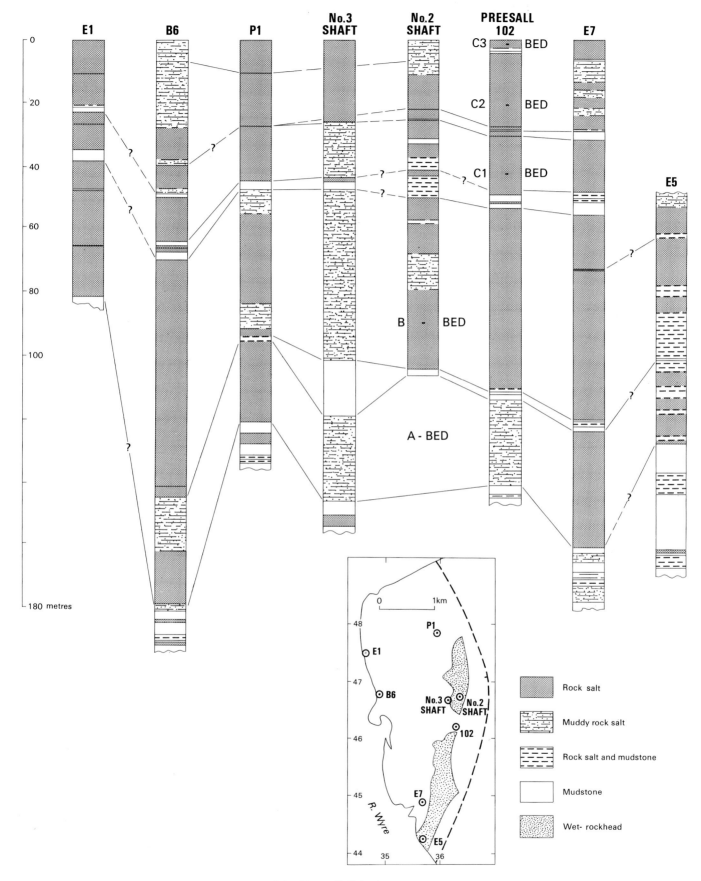

Figure 11 Representative vertical sections of the Preesall Salt

0   2   4cm

Plate 11   Collapse-breccia due to recent total solution of the Preesall Salt. At 70.70–70.90 m depth in the Thornton Cleveleys Borehole. (MN 26862)

the trough and along its axis, though there are local anomalies due to temporary uplift and contemporaneous solution along its axis. In such a setting, rapid thinning and deterioration in quality might be expected across the bounding structures, and the logs of boreholes E1 and E5 may record this effect.

It is, however, far from clear how far the down-faulted trough continued to the south-west, where three widely scattered boreholes, at Thornton Cleveleys, Churchtown and Mythop, all prove thick collapse-breccias in the stratigraphical position of the Preesall Salt (Plate 11). The breccias are up to 45 m thick and directly overlie the highest clastic cycle in the Thornton Mudstones. Their lithologies are more or less fragmented mudstones and siltstones, and in places the beds are buckled and steeply dipping. Gypsum porphyroblasts are common and there are bands of gypsum, up to 35 cm thick, towards the top of the breccias. Their presence suggests that the Preesall Salt was laid down well beyond its present limits in this direction. Indeed, some salt may still be preserved as isolated masses protruding up into the breccia, for there are possible subsidence hollows near Norbreck [316 411], Norcross [329 412], Little Carleton [324 382] and Mythop [359 352 and 365 351]. However, penecontemporaneous solution may well have been responsible for some of the breccias, for salt deposition within the down-faulted trough may have been contemporaneous with haselgebirge formation along its peripheries, and this latter lithology may have been dissolved away soon after its formation. In this view, the breccias are the lateral equivalents of the Preesall Salt rather than the fragmented residues of the overlying Coat Walls Mudstones.

COAT WALLS MUDSTONES

The Coat Walls Mudstones were completely penetrated by Coat Walls Borehole (Figure 12). This section now supplements the earlier, less detailed one from B6 Borehole. The member is 129.15 m thick (122.5 m when corrected for dip at Coat Walls), and comprises dominantly reddish brown mudstone interbedded with many bands of greenish grey mudstone generally less than 1 m thick. At Kirkham, the Coat Walls Mudstones were no more than 35 m thick. The greenish grey bands are usually much thinner than in the Thornton Mudstones and make up a smaller proportion of the total thickness. Towards the base of the sequence, just above the Preesall Salt, the greenish grey bands are thicker and more numerous,whilst bands of structureless reddish brown mudstone are less common. Much of the rest of the sequence is interlaminated with siltstone, but there are also many structureless mudstone interbeds which are invariably reddish brown. Towards the top of the sequence, these tend to be dominant, though laminated, greenish grey beds still occur. Bands of brecciated mudstone with some siltstone occur at several horizons in the sequence, but each is less than 1 m thick. They may be the lateral equivalents of thin beds of salt closer to the centre of the Manx–Furness Basin.

As in the Thornton Mudstones, the laminated beds contain many desiccation cracks, which were noted at 40 horizons in Coat Walls Borehole and were also common in the B6 Borehole. In addition, cross-lamination and current-ripple lamination occur in the siltstones. One of these, from Staynall Borehole [3562 4438], was analysed by X-ray dif-

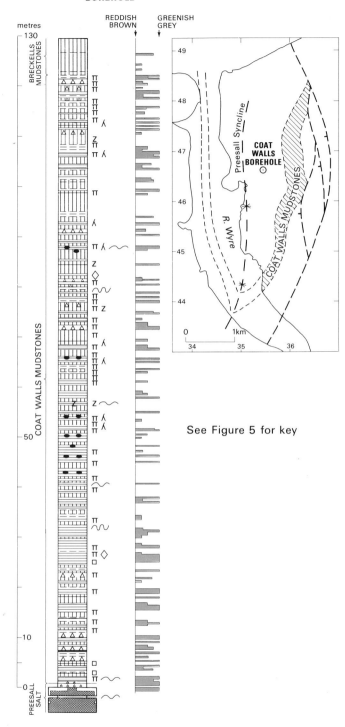

Figure 12 Borehole section in the Coat Walls Mudstones

fraction; it contained dolomite, quartz, illite and chlorite. Sporadic thin bands of mudstone with up to 20 per cent of halite crystals occur in the basal 27 m of the Coat Walls Mudstones (Figure 12).

Breckells Mudstones

The Breckells Mudstones have been proved in Coat Walls and Hackensall Hall boreholes (Figure 13), and in the earlier B6 Borehole. The proved sequences in the boreholes overlap and can be correlated using greenish grey bands as markers. This gives a total proved thickness of 209 m, but the top of the formation is not preserved in the present district.

The Breckells Mudstones can be divided into three lithological sequences. The lowest is 53 m thick and is dominantly reddish brown, structureless mudstone with scattered, greenish grey bands and patches that become more numerous towards the base. Thin, laminated or poorly laminated bands occur at widely spaced intervals. Gypsum nodules are virtually absent, though veins of gypsum are common (Plate 12). Halite is present in scattered thin bands, crystals and veins (Figure 13). In both the B6 and Coat Walls boreholes, bands of halite-gypsum, halite-anhydrite and halite, up to 22 cm in thickness, are present. They are commonly associated with breccias and are likely to be the attenuated remnants of thicker salts offshore, as for instance in the 110/3-2 Borehole (Jackson et al., 1987, fig. 9). Scattered halite veins occur in places, with gypsum selvages (Plate 13). In Coat Walls Borehole, any halite in the uppermost 30 m of these potentially salt-bearing beds has apparently been dissolved by groundwater. Halite is also present in a few veins and as halite crystals and seams in E1 and P1 boreholes, and in unspecified habit scattered through a thickness of 42 m in Preesall No. 8 Borehole [3536 4712], at an horizon comparable with the beds of salt in B6 and Coat Walls boreholes. Audley-Charles (1970, pl. 2) characterised about 30 m of the above strata in B6 Borehole as salt, but in reality there is only about 1 m of primary halite, or halite-gypsum, plus a number of halite veins, in these beds. He equated them with the Wilkesley Halite of Cheshire, but this seems unlikely in the light of later provings offshore.

The middle division, 91 m thick, consists of reddish brown, structureless mudstone with abundant nodules of botryoidal gypsum up to 5 cm across (Plate 14). The nodules commonly occur in bands and irregular concentrations, many being composed of myriads of small nodules interlinked by small gypsum veins. Gypsum veins are not ubiquitous, however, and do not occur in the highest 24 m of the succession. In Coat Walls Borehole, gypsum nodules are absent from the 33 m of strata immediately below rockhead but this is most probably due to leaching.

The highest lithological sequence in the Breckells Mudstones comprises 65 m of reddish brown, largely brecciated mudstone with a number of greenish grey patches and bands, especially near the top of the succession. These are the highest Triassic rocks known in the Fylde, though yet higher beds may be preserved in the centre of the Kirkham Syncline. Many of the brecciated mudstones are crudely banded and contain clasts generally less than 1 cm across, though more rarely up to 5 cm. Dips are anomalously high in the borehole cores; they range up to 45°, as compared with

0   2   4cm

Plate 13   Halite vein flanked by gypsum. The vein is associated with the basal part of the Breckells Mudstones (removed from the specimen) which also contain thin beds of halite. At 145.60–145.80 m depth in the Coat Walls Borehole. (MLD 10126)

0   2   4cm

Plate 12   Structureless reddish brown mudstone with gypsum veins, in the Breckells Mudstones. An earlier low-angled system of veins is split by a later, thicker, high-angled vein. At 120.35–120.50 m depth in the Coat Walls Borehole. (MLD 10128)

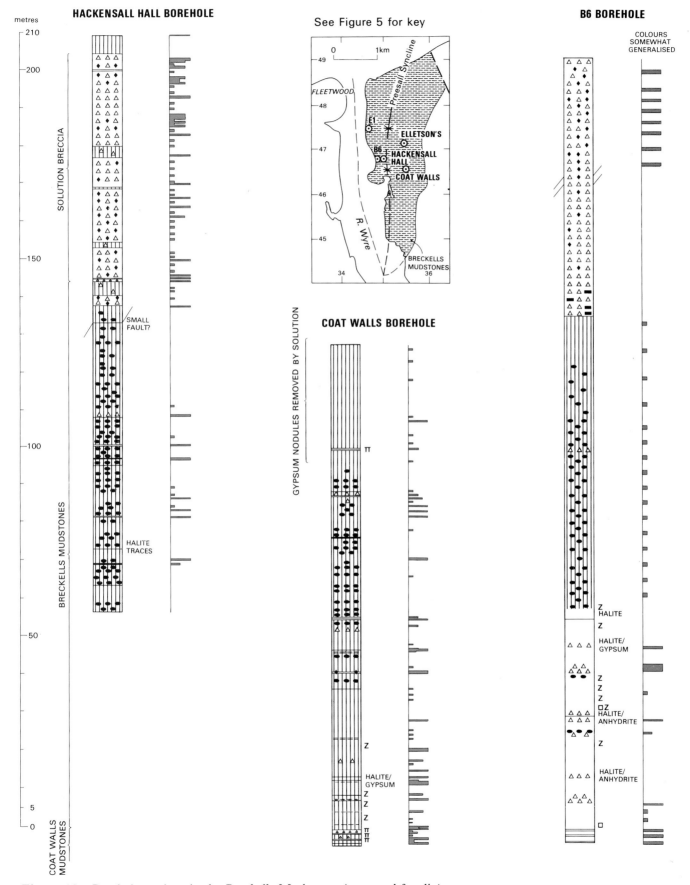

Figure 13 Borehole sections in the Breckells Mudstones (corrected for dip)

0   2   4 cm

Plate 14 Gypsum nodules and veins in structureless reddish brown mudstone, in the Breckells Mudstones. From 147.15–147.34 depth in the Hackensall Hall Borehole. (MLD 10127)

the regional dips of about 8°, suggesting local collapse due to the solution of evaporites. Four horizons of intense brecciation can be recognised in Hackensall Hall Borehole and are separated by strata that are only slightly brecciated. Prior to solution, these may have been four discrete beds of rock-salt or haselgebirge, separated by mudstones with little halite. The sequence may correlate with 550 m of halite proved offshore in the East Irish Sea Basin (Jackson et al., 1987). If so, then this highest division of the Breckells Mudstones could well be designated as a formation in its own right.

Above these breccias in Hackensall Hall Borehole are 4.82 m of reddish brown, structureless mudstone containing some gypsum porphyroblasts in the basal 1.45 m. Though these beds have not obviously collapsed, it is possible that the underlying breccias are produced by recent rather than Triassic solution and collapse, since they are rich in gypsum porphyroblasts like many recent salt solution-breccias. If so, these highest mudstones probably form a raft, let down *en bloc* by the underlying collapse. As in B6 Borehole, the highest beds in the solid section of this borehole were not cored, so that a critical part of the evidence for the age of the collapse is missing.

CONDITIONS OF SEDIMENTATION

The conditions under which the Triassic rocks were laid down are not easy to envisage, partly because there seem to be no precise modern analogues. The red sediments, aeolian sands and salt deposits have commonly been taken to imply deposition in a land-locked desert, studded by ephemeral lakes and lying far from marine influences.

Within the present district, too few details are available from the Sherwood Sandstone Group to test the validity of this model. In parts of the succession, the sandstones are mainly composed of aeolian grains, but whether these are all fossilised dune-fields is less certain, for the cores do not closely resemble other aeolian sandstones, such as the Collyhurst Sandstone or the Frodsham Beds. Other parts of the Sherwood Sandstone are certainly water-laid. Indeed, in north-western England as a whole, fluvial sandstones seem to make up most of the succession. How much is fluvial of the 1700 m of sandstone recorded in the northern part of the East Irish Sea Basin (Jackson et al., 1987) is uncertain, but southwards in Cheshire much of the 1000 m or so of the Sherwood Sandstone is now regarded as the product of rivers (Thompson, 1970a and b). It is not easy to assess how far alluvial sequences of such thicknesses are compatible with the desert environment that is traditionally envisaged. Nor is the climatic significance of the widely distributed aeolian grains certain, for the grasses that now produce a continuous soil cover in humid climates had not then evolved, and it is likely that sand was far more mobile when there was no such cover. It seems probable that sedimentation occured across a wide plain with low, longitudinal gradients, crossed by migrating rivers that had already largely dropped the coarser components of their bed-load. The rivers were probably separated by emergent tracts on which sand-dunes accumulated periodically. Clearly sedimentation and subsidence had to proceed hand-in-hand, to allow the same limited range of facies to persist through such a considerable thickness of strata.

The Mercia Mudstone Group yields much more sedimentary data, though little relates to the thicker halites. Four main facies are involved:

A Structureless, reddish brown mudstone with gypsum nodules and veins

B Interlaminated, grey and pale brown mudstones and siltstones with gypsum veins, current-ripple laminations, pseudomorphs after halite, and desiccation cracks

C 'Haselgebirge' lithologies, with halite crystals set within a matrix of red mudstone

D Bedded salt with thin partings of mudstone.

Facies A seems to be largely loessic in origin (Earp and Taylor, 1986) and the lack of bedding may be due to the penecontemporaneous mixing and resettling of sediment before final consolidation. Facies B was laid down in shallow water that periodically dried out. In Facies C, the rock-salt is likely to have been derived from saline ground-water within emergent sediments (Shearman,1970). Finally, Facies D ap-

pears to represent primary accumulations of halite beneath a shallow body of water that occasionally dried-out, only to reform at the same site. The various formations of the Mercia Mudstone are characterised by different combinations of the above facies, and by the different thicknesses of each that are developed. The difficulties in interpretation largely arise from the innumerable changes of facies, and what these mean in terms of depositional geomorphology.

The Hambleton Mudstones are dominated by Facies B, to the virtual exclusion of the others. Similar lithologies in the almost contemporaneous 'Waterstones' of north Cheshire have been claimed to be indicative of tidal flats (Ireland et al., 1978), and Earp and Taylor (1986) consider that the base of the 'Waterstones' marks a major marine transgression. The presence of organic-walled microplankton (acritarchs) in the Hambleton, Singleton and Kirkham mudstones of the present district (Figure 14) supports this belief, as they do not occur in the loessic lithologies of Facies A and were not, therefore, introduced by wind transport. The sea,

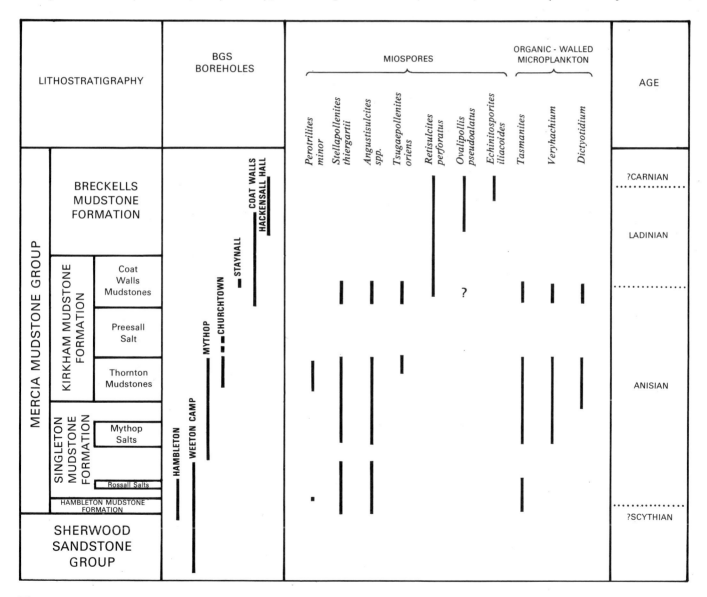

Figure 14 Occurrences of selected palynomorphs in the Mercia Mudstones

however, must have advanced over a flat coastal plain of great extent, flooding it with very shallow water and retreating repeatedly to expose broad mud-flats. The evidence suggests a coast-line quite unlike modern ones, particularly in the width of the coastal plain subject to flooding.

The succeeding Singleton Mudstones are largely composed of Facies A and some Facies B, with local developments of Facies C and even Facies D. Presumably, by this time, the sea had retreated allowing sabkha conditions to set in over the whole coastal plain, relieved only by local saline pools where bedded salt was deposited.

The Thornton Mudstones mark a return to the depositional conditions of Facies B. In the Coat Walls Mudstones however, Facies B alternates cyclically with Facies A. It is far from clear that this alternation of facies represents regular, short-lived, marine transgressions and regressions. If it does, however, then the sea was too saline to support marine faunas and consequently some authors prefer the view that the sequence was laid down in a lake-basin that regularly dried-out. Acritarchs reappear, in both the Thornton and Coat Walls successions, strengthening the belief in a coastal location for Facies B.

Set between the Thornton and Coat Walls sequences, the Preesall Salt is almost wholly of Facies D, and is typical of the deposits traditionally assigned to desert lakes in this part of the sequence. It is unusual, however, for such lakes to deposit practically nothing but halite, while the association with acritarch-bearing strata suggests the proximity of the sea. Perhaps brine accumulated within an occasionally inundated coastal salina; alternatively it may have become concentrated within an actively sinking area of the coastal plain which was regularly fed by the sea, although on at least two occasions it is known to have dried out.

Above the Coat Walls Mudstone, there was a lessening of the marine (or lacustrine) influence and the Breckells Mudstone reverts mainly to the lithologies of Facies A, like those of the earlier Singleton Mudstones. Thus, the collapse-breccias at the top of the sequence resemble the salts in the Singleton Mudstones in not being associated with the development of Facies B lithologies. Nevertheless, the breccias appear to correlate with thick salt-beds known beneath the Irish Sea, so they probably mark a return to rock-salt deposition.

PALYNOLOGY

Samples from the uppermost 55 m of the Sherwood Sandstone succession in Weeton Camp Borehole (Figure 4) yielded sparse organic residues containing only a few indeterminate or vestigial palynomorphs affording no biostratigraphical evidence of age.

Productive palynological samples have been recovered from most of the formations in the Mercia Mudstone Group from the Hambleton Mudstones to the Breckells Mudstones. Rich and varied assemblages of palynomorphs recovered from Hambleton, Mythop and Churchtown boreholes, (Figures 5, 6, 7 and Plate 15), representing most of the succession below the Preesall Salt, and from Staynall Borehole, representing a short section within the Coat Walls Mudstones, were reported by Warrington (1974a). Subsequent examination of material from Coat Walls and Hackensall Hall boreholes (Figures 12 and 13) has completed the palynological coverage of the succession above the Preesall Salt. Lower in the sequence, there was previously a gap in the palynological sampling of the Singleton Mudstones between sections in their lower and upper parts proved, respectively, in Hambleton and Mythop boreholes, but this gap has now been partly filled by samples from Weeton Camp Borehole. Warrington (1974a) gave no dating of the sequence in terms of the standard Triassic stages, though potential biostratigraphical subdivisions of the local succession were identified; provisional ages based upon the palynological results were, however, applied by Warrington to the succession on the Blackpool sheet (1975). The results now available from the incomplete Mercia Mudstone succession preserved within the district permit a revision of these earlier datings and correlations, particularly in the light of recent palynological studies of Triassic sequences in southern Europe where independent fossil evidence of age is available.

In Coat Walls Borehole, there is evidence for the position of the Anisian–Ladinian stage boundary. The miospore *Stellapollenites thiergartii* (Mädler) Clement-Westerhof et al., 1974, which Visscher and Brugman (1981) and van der Eem (1983) show ranging only to the top of the Anisian Stage, has been recorded in the Coat Walls Mudstone up to about 17 m above the Preesall Salt. Slightly higher in the borehole, miospores indicative of a Ladinian age appear, including *Retisulcites perforatus* (Mädler) Scheuring 1970 about 20 m above the Preesall Salt, and possible specimens of *Ovalipollis pseudoalatus* (Thiergart) Schuurman 1976 about 8 m higher. Definite occurrences of *O. pseudoalatus* are known from the Breckells Mudstones higher in Coat Walls Borehole, and from Hackensall Hall Borehole where they are associated with *R. perforatus* and *Echinitosporites iliacoides* Schulz and Krutzsch 1961. It is not possible to determine satisfactorily whether the top of the Ladinian Stage, at its boundary with the Carnian, occurs within the sequence studied; *E. iliacoides* and *R. perforatus* range into the early Carnian (Cordevolian substage), according to Mostler and Scheuring (1974), and *O. pseudoalatus* is present throughout late Triassic sequences. In Hackensall Hall Borehole, *E. iliacoides* and *R. perforatus* are more prominent in assemblages from lower in the section and *O. pseudoalatus* is more evident in the higher assemblages, features which suggest, by comparison with Scheuring's studies (1970), that beds in the higher part of that section may be of Carnian age.

The Mercia Mudstone succession up to about the top of the Preesall Salt was previously assigned to the Scythian and Anisian stages (Blackpool sheet, 1975) on the basis of a comparison of Warrington's (1974a) palynological results with those available from dated Triassic sequences in southern Europe. Comparison of the Blackpool records with more recent work on independently dated Triassic sequences indicates that the succession from the Hambleton Mudstones up to that identified here in the basal Coat Walls Mudstones as the Anisian–Ladinian boundary, is largely, if not entirely, of Anisian age. This is suggested by the presence of taxa such as *Angustisulcites spp.* and *Stellapollenites thiergartii*, which Visscher and Brugman (1981) and Brugman (1986) document only from Anisian sequences. These taxa occur in

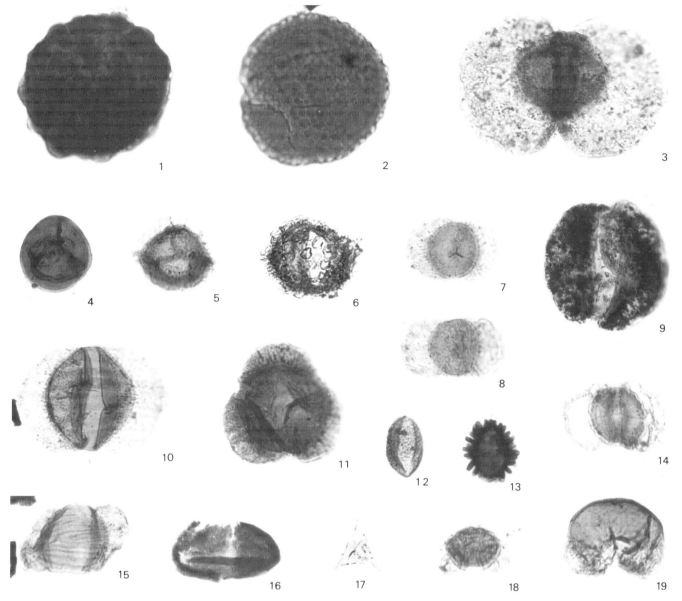

Plate 15 Palynomorphs from the Mercia Mudstones (magnification ×500)

1 *Verrucosisporites thuringiacus* Mädler 1964. MPK 6418. Hambleton Mudstones; Hambleton Borehole, 135.79 m
2 *Verrucosisporites jenensis* Reinhardt & Schmitz *in* Reinhardt 1964. MPK 6419. Hambleton Mudstones; Hambleton Borehole, 135.79 m
3 *Angustisulcites grandis* (Freudenthal) Visscher 1966. MPK 6420. Kirkham Mudstones; Mythop Borehole, 93.65 m
4 *Densoisporites nejburgii* (Schulz) Balme 1970. MPK 6421. Kirkham Mudstones; Churchtown Borehole, 78.44 m
5 *Perotrilites minor* (Mädler) Antonescu & Taugourdeau Lantz 1973. MPK 6422. Hambleton Mudstones; Hambleton Borehole, 108.38 m
6 *Tsugaepollenites oriens* Klaus 1964. MPK 6423. Kirkham Mudstones; Staynall Borehole, 44.67 m
7 *Triadispora crassa* Klaus 1964. MPK 6424. Kirkham Mudstones; Coat Walls Borehole, 171.4–171.55 m
8 *Triadispora plicata* Klaus 1964. MPK 6425. Kirkham Mudstones; Coat Walls Borehole, 171.4–171.55 m
9 *Angustisulcites gorpii* Visscher 1966. MPK 6426. Hambleton Mudstones; Hambleton Borehole, 108.38 m

10 *Lunatisporites puntii* Visscher 1966. MPK 6427. Kirkham Mudstones; Churchtown Borehole, 92.58 m
11 *Stellapollenites thiergartii* (Mädler) Clement-Westerhof, van der Eem, van Erve, Klasen, Schuurman & Visscher 1974. MPK 6428. Hambleton Mudstones; Hambleton Borehole, 108.38 m
12 *Retisulcites perforatus* (Mädler) Scheuring 1970. MPK 6429. Breckells Mudstones; Hackensall Hall Borehole, 94.0–94.2 m
13 *Echinitosporites iliacoides* Schulz & Krutzsch 1961. MPK 6430. Breckells Mudstones; Hackensall Hall Borehole, 94.0–94.2 m
14 *Angustisulcites klausii* Freudenthal 1964. MPK 6431. Singleton Mudstones; Mythop Borehole, 246.35 m
15 *Striatoabieites balmei* Klaus emend. Scheuring 1978. MPK 6432. Hambleton Mudstones; Hambleton Borehole, 128.50 m
16 *Ovalipollis pseudoalatus* (Thiergart) Schuurmann 1976. MPK 6433. Breckells Mudstones; Hackensall Hall Borehole, 46.80 m
17 *Veryhachium* sp. MPK 6434. Kirkham Mudstones; Staynall Borehole, 44.67 m
18 *Protodiploxypinus doubingeri* (Klaus) Warrington 1974. MPK 6435. Singleton Mudstones; Mythop Borehole, 205.99 m
19 *Protodiploxypinus fastidiosus* (Jansonius) Warrington 1974. MPK 6436. Kirkham Mudstones; Churchtown Borehole, 92.58 m

assemblages from the Hambleton Mudstones in Hambleton Borehole, from the Singleton Mudstones in that section and from Weeton Camp and Mythop boreholes, and the Thornton Mudstones in the Mythop and Churchtown sections.

In general terms, the succession above the Preesall Salt correlates with the Lettenkohle and lower Gipskeuper in the Trias of Germany, and lower beds in the Mercia Mudstone correlate with the German Muschelkalk and the underlying Röt or upper Bunter. More local correlations are indicated by certain miospores recorded in the Blackpool succession. The taxa *Echinitosporites iliacoides* and *Retisulcites perforatus*, that are known from the Coat Walls and Breckells mudstones (Figure 14), have also been recorded from a level high in the Glenstaghey Formation in the Mercia Mudstone in Northern Ireland (Warrington, 1978a), and from the Carlton Formation to basal Edwalton Formation succession in the Mercia Mudstone of central and eastern England (Warrington, 1970a; Smith and Warrington, 1971; Fisher, 1972; Warrington, 1988). Records of *Tsugaepollenites oriens* Klaus 1964 from the succession from the upper part of the Thornton Mudstones in Mythop and Churchtown boreholes to the Coat Walls Mudstones in Coat Walls Borehole (Figure 14) indicate correlations with beds in the lower part of the Glenstaghey Formation in Port More Borehole, Northern Ireland (Warrington, 1978a) and in the Mercia Mudstone between 993 and 1070 m in Langford Lodge Borehole, Northern Ireland (Warrington, 1970b); other occurrences of *T. oriens* are in mudstones overlying the Northwich Halite in the Mercia Mudstone in Cheshire (Warrington, 1986), in the Denstone Formation and succeeding beds in the lower part of the Mercia Mudstone in the Ashbourne district (Warrington, 1982), in equivalent beds in the adjoining Derby district (Warrington *in* Frost and Smart, 1979), and in the upper part of the Sherwood Sandstone Group near Stratford-upon-Avon (Warrington, 1974b) and possibly near Banbury, Oxfordshire (Warrington, 1978b). In Cheshire and the Ashbourne and Derby districts *T. oriens* has been found in association with *Perotrilites minor* (Mädler) Antonescu and Taugourdeau-Lantz, 1973, which has also been observed in the Thornton Mudstones from Churchtown and Mythop boreholes in the Blackpool district (Figure 14). The occurrences of *T. oriens* indicate a general correlation between the Preesall Salt of west Lancashire and the Northwich Halite in Cheshire.

Organic-walled microplankton have been recorded in small numbers throughout much of the Mercia Mudstone succession in the Blackpool district (Figure 14). Prasinophyte algae (*Tasmanites*) are present in the assemblages from the Hambleton Mudstones to the Coat Walls Mudstones. Acritarchs, including representatives of the polygonomorphitae (*Veryhachium*) and herkomorphitae (*Dictyotidium*), have been recovered from the Thornton and Coat Walls mudstones. The presence of these remains is indicative of deposition in an aqueous environment connected to a marine source and favours the concept of derivation of the substantial bodies of halite present in the sequence from marine brines (Warrington, 1974c).

The palynological dating of the Mercia Mudstone in the Blackpool district as largely or entirely Anisian and Ladinian permits an assessment of the rate of subsidence and sedimentation in the district during deposition of the Hambleton Mudstone to Breckells Mudstone succession. This succession, more than 800 m thick, represents a period of time of approximately 12 million years, between 242 ± 5 and 230 ± 5 Ma (Forster and Warrington, 1985).

CHAPTER 3

Structure

The small number of deep boreholes within the district means that the detailed structure is still largely conjectural. Meanwhile, the current interpretation of the structure mainly invokes folding rather than faulting. Both processes are almost certainly involved but it may be that growth faulting along the margins of the depositional basins is the most important element of the structure. Conjectural structure contours on the base of the Mercia Mudstones are shown in Figure 15. There are two major synclines, the Preesall Syncline in the north and the subparallel Kirkham Syncline just outside the district to the south-east, which are separated by three smaller folds.

The Preesall Syncline is thought to trend roughly north–south with its axis close to the eastern bank of the Wyre estuary. Boreholes in the saltfield on its eastern limb prove an average westward dip of about 15°. To the east, the dips generally terminate against the Preesall Fault-zone (Figure 15) which has at least two component fractures that have the overall effect of throwing Mercia Mudstone against Sherwood Sandstone; the boreholes are too widely spaced to plot the course of the fractures with certainty. Locally, the dip turns over to the east near the bounding faults, especially near Park Cottage [3632 4601] where a minor north-trending anticline is aligned parallel to the faults on their downthrow side (Figure 8, section c–c1). To the west, there is still considerable doubt about the structural interpretation. An overall eastward dip of about 25° along the eastern bank of the Wyre is suggested by two of the three westernmost boreholes in the saltfield and this dip, if it continues, could account for the known distribution of strata, which would then lie in a half-graben. Specific dip records from the boreholes are, however, scanty. Dips of about 25° are recorded below 300 m in B6 and E2 boreholes, but at shallower depths and in E1 borehole much gentler dips obtain. It is thus possible that the general dip is shallow and that there are down-east faults antithetic to the Preesall Fault along the western margin of the syncline. In either view, there is a rapid flattening of the dip west of the Wyre. Although information is sparse, it seems likely that the fold plunges to the north, and opens out towards the coast (Figure 15). To the south, a swing in the strike near Burrow's Marsh [354 451] suggests that the syncline is closing, but the presence of possible subsidence hollows (p.24) can be interpreted as indicating a shallow, southward extension of the fold.

Farther south, Churchtown and Mythop boreholes prove further breccias roughly at the stratigraphical horizon of the Preesall Salt, suggesting that the Preesall Syncline may continue southwards as a broad, open fold (Figure 15).

The broad outline of the Kirkham Syncline has been determined by drilling in the Garstang district to the east. There appear to be two separate axes, a western one trending through Elswick and another extending from Inskip to a little west of Kirkham. The eastern limb of this latter fold was penetrated by Kirkham Borehole, which proved average dips of 28°, exceeding those on the eastern limb of the Preesall Syncline. The town of Kirkham stands on a sizeable gravity anomaly, possibly caused by a rapid and very considerable local thickening of the Sherwood Sandstone (Barker, 1974), or by an underlying Permian basin. The core of the syncline has not been drilled, and it is possible that a small Jurassic outlier may be preserved within it.

These two synclines are separated by the Weeton Anticline. Along the axis of the fold, Weeton Camp Borehole proved the top of the Sherwood Sandstone at 134 m below OD. Another borehole only 4 km to the north reached the same horizon at 61 m below OD, suggesting that the fold has a gentle plunge to the south-west. Dips in both boreholes were about 6°.

Two subsidiary folds appear to separate the Weeton Anticline from the Preesall Syncline. The western one is an anticline trending through Poulton-le-Fylde; its presence is suggested by a borehole north-east of the town which is said to have entered the Sherwood Sandstone at 140 m below OD, appreciably higher than in B4 and B5 boreholes. These latter boreholes seem to lie within a gentle syncline trending towards Hambleton. The south-western continuation of this syncline is uncertain but it is possible that it merges into a shallow, southerly extension of the Preesall Syncline.

The tract extending across the district from the outskirts of Blackpool through Warbreck and on to Cleveleys and Fleetwood seems to be characterised by gentle landward dips, though the evidence for this is scarce.

The trends of the postulated folds and faults of the present district, which generally lie between north and north-east, are subparallel to those of several folds in the Namurian rocks near Lancaster. Faulting along the same general trend is well known to the north in Furness, and farther south in the Lancashire coalfield; the trend contrasts markedly with the west-south-west structures of the Ribblesdale Fold Belt. More significantly, the trend is followed by most of the faults between the Fylde coast and the off-shore gasfield (Jackson et al., 1987 and Figure 1).

The age of these structures has long been debated. Where only Triassic rocks are involved, the structures were once held to be Tertiary in age. It is now clear, however, that many of them profoundly influenced Permian and Triassic sedimentation and so were active during those periods; it seems likely that they were activated by the east–west tensional stresses which eventually led to the opening of the Atlantic in Cretaceous times. In the Lancashire coalfield, however, some faults sharply increase their throw where they pass downwards from the Permian cover into the Westphalian rocks (Jones et al., 1938), showing that they are much older fractures which were re-activated by Permo-Triassic stresses. Indeed, there were highly active sedimentary hinges in late-Dinantian times in southern Furness (Rose and Dunham, 1977) which utilised end-Silurian faults. It seems likely, therefore, that many of these structures lie on inherited end-Silurian lines which were intermittently active at least until Jurassic times.

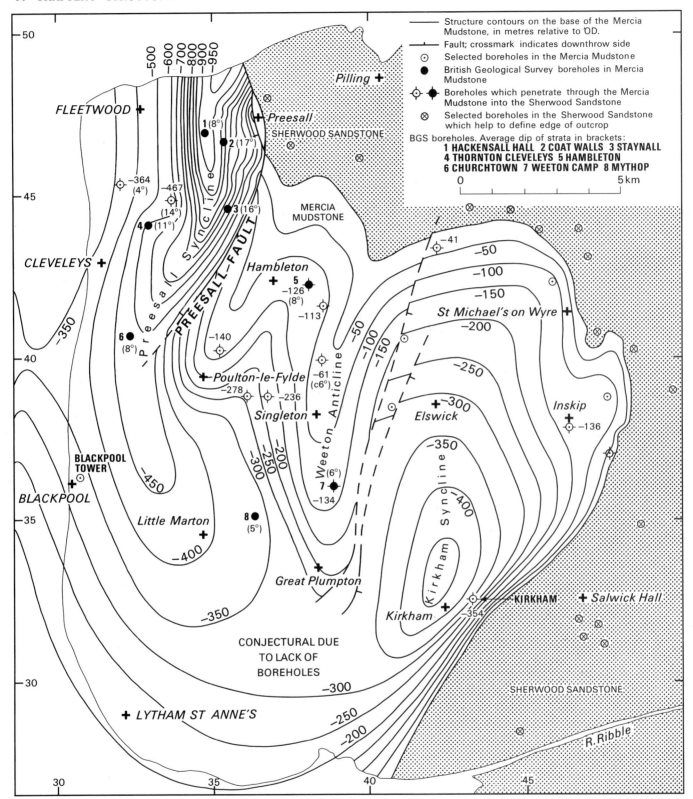

Figure 15   Structure contours on the base of the Mercia Mudstones

Since this memoir went to press, newly acquired borehole and seismic data from country east of the Blackpool district allows a more detailed structural analysis of the Kirkham Syncline. The updated structural picture with associated faults and flexures will be illustrated in the forthcoming memoir on the Garstang district

CHAPTER 4
Glacial deposits

Rockhead

Throughout the entire district the Triassic rocks are overlain by a cover of Quaternary deposits and, since rockhead is probably everywhere below Ordnance Datum, only the drift accumulations bring the western Fylde above sea-level. Though there are many boreholes into the Quaternary deposits only a few scattered provings reach rockhead, except within the Preesall saltfield where rockhead is known with much greater precision. Two considerations limit the use of borehole data in contouring the rockhead surface. Firstly, the methods of drilling and sampling generally employed make it difficult to distinguish lithologically between Boulder Clay and Mercia Mudstone and, in the east, between Glacial Sand and the Triassic sandstones, particularly where casing intervals have allowed cavings to develop. The depths to rockhead recorded in driller's logs are thus commonly unreliable. Secondly, to draw contours between widely spaced boreholes requires some concept of the form of the surface being contoured. The amount of glacial modification to the preglacial surface is not known, however, nor is it certain whether the surface was subaerial or submarine.

The recorded point-data are summarised in Figure 16 which also, despite the above reservations, attempts to provide rockhead contours. Recorded rockhead varies from 61 m below OD [3090 3621] at the Winter Gardens, Blackpool, to 3 m below OD [3397 4480] alongside the Wyre north of Burn Naze. Within the Preesall saltfield, where the data are most abundant, there are rapid and substantial variations in the recorded depths, with a complex, ungraded depression crossing the tract from north to south; had the borehole density been as low at Preesall as it is elsewhere, these variations would not have been detected. It also seems clear that a topographical hollow in the predrift surface overlies the wet-rockhead area, and is now concealed by the glacial deposits (Figure 8). Other groupings of data-points, such as those along the Wyre between Staynall and Burn Naze, near Fleetwood and Knott End-on-Sea, and north of Primrose Bank [332 415], show fluctuations of 15 to 20 m between closely adjoining boreholes, while borings for the sewer outfall on the Manchester Square foreshore in Blackpool show that the local rockhead varies from 21 m below OD to below 31 m below OD within a horizontal distance of 400 m. Local gradients on the rockhead surface are clearly greater than any regional pattern that can be imposed on the data, and this emphasises that the contours shown in Figure 16 are not likely to be reliable in forecasting the depth of rockhead at any particular point. It does seem certain, however, that the glacial deposits do not rest on a gently sloping, wave-cut platform.

The ubiquitous glacial deposits are almost certainly the products of only one glaciation, and they are generally all regarded as of late-Devensian age, which is certainly the age of the surface till; they were probably laid down between

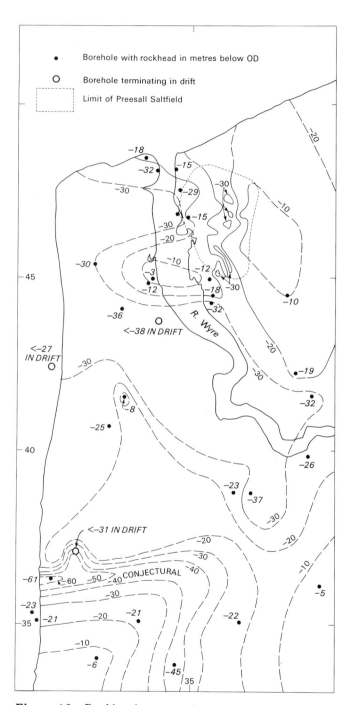

Figure 16 Rockhead contours in metres.

These are more firmly established in the saltfield than elsewhere

20 000 and 12 000 years BP.

Where these deposits crop out at the surface, they give rise to three distinctive types of terrain, shown in Figure 17. In Area A, the glacial sequence is the three-fold one typical of many parts of north-west England, with a compact, basal till separated by thick and laterally continuous, water-laid sands from an upper till. In Area B, the sands are generally absent and the glacial deposits mostly comprise drumlins of till, some with cores of sand. Area C is largely composed of sheets of till but no drumlins; there are several extensive flats that mark the former sites of glacial lakes. Each of these areas is described separately below for each has its own particular geological problems.

AREA A—BLACKPOOL TO WEETON

The broad belt of comparatively high ground comprising Area A (Figure 17) extends westwards across the southern Fylde from the northern outskirts of Preston to the coast at Blackpool. Within the present district it is 5 to 8 km wide, rising to 40 m above OD east of Weeton. Included for convenience within Area A is a small, lower-lying tract south of Central Pier, Blackpool.

Surface exposures are generally poor, except for the cliff-sections southwards from Norbreck, and even these are far less clear than they were in the last century (Binney, 1852; De Rance, 1877). Many site investigation boreholes have been drilled in Blackpool, however, mostly to depths of 10 to 30 m. More than 3000 records are held by Blackpool Borough Council and by BGS and, farther east, there are more boreholes sunk to investigate the line of the M55 Motorway. Exposures and boreholes show that the tract is characterised by two tills of strikingly different properties, separated mainly by sand, though in detail there are thin, local bands of laminated clay, silt, sand and gravel within the tills, and lenses of silt, laminated clay and till within the sand. Boreholes show that the distribution of the higher ground and of minor topographical features along it owes much to variations in the thickness of the sand.

The broadly tripartite nature of the drift was recognised by both Binney and De Rance, who applied the terms Lower Boulder Clay, Middle Drift (or Sand/s) and Upper Boulder Clay. Though currently unfashionable, and criticised by Longworth (in Johnson, 1985), these remain useful descriptive subdivisions and have been retained in the present account, though their usage should not be taken to imply strict contemporaneity with deposits that have been similarly named farther south in Lancashire and in Cheshire.

Lower Boulder Clay

The best sections in the Lower Boulder Clay remain those recorded by Binney and De Rance in the coastal cliffs (Figure 18). These authors record that the till arches up in several points along the section. The northern rise near Red Bank Road [about 3075 3973] shows 7.6 m of till, and the southern one near Uncle Tom's Cabin [about 3042 3858] shows 2.7 m of 'indian red' till (De Rance, 1877); the colour of the till possibly indicates that solid Mercia Mudstone is not far below, though the sparse borehole data hereabouts

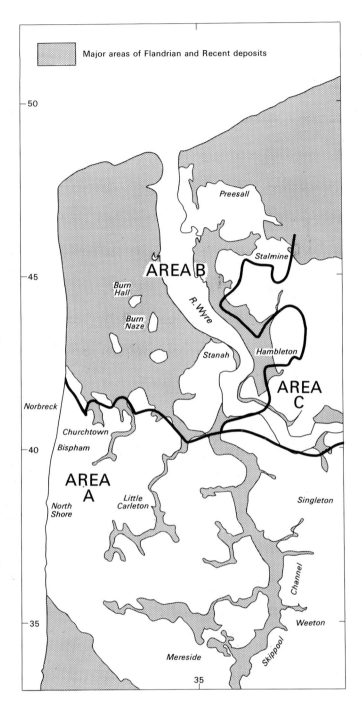

Figure 17   Descriptive areas of Devensian deposits

suggests that rockhead is at 20 m below OD. Northwards, the upper surface of the till falls to below beach-level, and a spring-line on the foreshore [3080 4025 to 3088 4077], along which lower till is exposed from time to time, marks its probable position.

Where the till is thickest it has been described as a gritty, compact clay rich in erratics, the largest recorded being almost 3 m across. It is devoid of internal bedding except for some layering of the included clasts near the top, and for a single plane near the base of the cliffs, below which rounded

and scratched stones are especially abundant. There is no
recorded sign of the weathering along this plane which might
have suggested that it is a junction between tills of different
glaciations. The deposit is everywhere extremely compact
and rich in erratics, largely of Cumbrian provenance. Binney (1852, p.130) noted abundant marine shells, and also
pieces of flint from the Chalk and the derived Jurassic fossil,
Gryphaea incurva. The most characteristic feature of the till is
its compactness and high consolidation. During construction
in 1917–21 of the shore defences that now almost totally
conceal the outcrop, piling was attempted without success.
The basal till was said to be of a 'rock-like toughness', and it
had to be blasted out of the trench on which the sea-wall is
founded (Banks *in* Grime, 1936). Some 65 years later, when
a continuous curtain of steel piles was installed at the base of
the sea-wall, it was in places only possible to use 2 m piles
because of the overconsolidated nature of the deposit.

Parallel to, and slightly inland from the coastal section,
borehole logs supplement these descriptions and extend
knowledge of the sequence southwards. Boreholes sunk on
the foreshore south of Central Pier appear to have encountered another local rise of lower till which reaches to
about 15 m below OD (Figure 19). The till is 2.1 m to at least
7.3 m thick and is described as a very hard, sandy and stony
clay. Towards the southern margin of the district, the surface of the lower till falls to about 28 m below OD, on the
southern flank of the Blackpool–Weeton ridge.

Inland, scattered boreholes confirm that rockhead is invariably overlain by stony clay but little is known of the
mechanical properties of the deposit. A recent hole [3888
3603] at Weeton Camp proved more than 18 m of brown,
sticky clay with pebble-sized erratics which does not appear
to be as compact as the lower till of the cliffs. There are,
however, no surface exposures to confirm this, the only
possible outcrop of the till being in the bottom of a flooded
sand-pit [3755 3577] south-south-west of Preese Hall Farm.

The best inland provings of the lower till are in a line of
boreholes drilled along the line of the M55. A small-scale section of these provings has been published by Longworth (*in*
Johnson, 1985), and the data have been plotted on a larger
scale in Figure 20. Within the district, few holes have
reached the underlying solid, but several have proved tills
that appear broadly to correlate with the lower till in the
coast sections. These tills may underlie the fill of the Skippool Channel (Figure 20), but nowhere else do they approach the surface.

Glacial Sand (Middle Sand)

The most extensive sections of Glacial Sand are in the
Blackpool cliffs where De Rance (1877) described 7.75 to
20 m of sand lying between the Lower Boulder Clay and the
surface till. Fortunately, De Rance recorded graphic sections
of the cliffs in his Notebook 2, stored in the BGS archives,
and these sections are reproduced for the first time with some
interpretation based on his field slips and memoirs (De
Rance, 1877). The sections have been largely obscured by
repeated coastal defence works and, more recently, by landscaping. Inland, there are scattered exposures of sand at this
stratigraphical horizon, noticeably around Weeton in the extreme south-east. A wealth of borehole data at Blackpool

shows that the Glacial Sand forms an almost continuous bed
of variable thickness hereabouts. Several of the deeper
boreholes prove an uninterrupted sequence of sand, but
some prove thin intercalations of clay within the sequence, as
at Marton Gasworks [3449 3369] where two layers of sandy
clay and one of till lie within an unusually thick sandy succession (Figure 20).

Boreholes sunk along the line of the M55 provide a section
along the length of the ridge (Figure 20), complementing the
transverse section given by the coastal boreholes at
Blackpool. They prove widespread, but locally discontinuous, sands of variable thickness beneath the surface tills;
there is a clear discontinuity between the sands on either side
of the Skippool Channel. Directly to the east of the district,
Longworth (*in* Johnson, 1985) records two persistent sands
separated by till in the M55 boreholes, but this sequence
does not seem to continue westwards to Weeton.

DETAILS

The coastal exposures were described in some detail by De Rance
(1877, pp.21–25). He noted that the basal 0.15 to 1.75 m of the
sands was a laminated loam resting conformably on rises of Lower
Boulder Clay. The overlying beds, 7.60 to 18.30 m thick, over-
lapped against the loam. They were mainly sands, with beds of
shingle and laminated loam and were especially well exposed near
Uncle Tom's Cabin [307 385] (De Rance, 1877, figs. 4,7), though
this part of the section is now entirely obscured. Current bedding in
the shingle was inclined towards the south-south-east. Marine shells
were found in the shingle and sand by Binney (1852), Mackintosh
(1869) and De Rance (1877). Hitherto unpublished details of the
lenses of silt and gravel within the sand are shown in Figure 20.

The sands are now largely obscured by concrete sea-defences
along most of the section, but at the time of the resurvey, several
gully sections were available near the cliff-top [3075 3972 to 3067
3893]. Only the topmost 4.60 m of commonly cross-bedded, fine-
grained sand containing a few pebbly bands were exposed. The topmost 0.05 to 0.70 m of sand were, in places, lithified with a
calcareous cement to give a rock that weathers into characteristic
fretted forms (Figure 18, illustration 3). Blocks of identical lithology
can be seen on the beach; the largest, the Pennystone [3045 4004],
is so large that De Rance thought that it must lie close to the position where it fell from the old cliff-line, about 400 m west of the present cliff (De Rance, 1877, fig.8). Sketches of its sphinx-like profile
made by De Rance in his field notebook a century ago show that the
block has changed little in shape during the intervening time. It
consists of a mass of well-cemented, coarse-grained sand, 2.4 m
high, with erratic pebbles and cobbles. Further calcreted stones of
similar lithology, but only 1 m high, stand close to the low-water
mark of spring tides [3021 4041]. Towards the north end of the cliff-
section, the presence of sand beneath till is obvious, but there are
small slips and flows along the cliff obscuring the precise relationships. Northwards from Norbreck Hydro [311 407], the cliffs seem
wholly of till capped by Blown Sand, though it is possible that the
sands are present beneath the slipped cliff-face. They almost certainly thin rapidly northwards, however, for De Rance recorded 0.6
to 0.8 m of cemented gravel on the cliff [3110 4164] at Little
Bispham, which is reminiscent of the top of the sands farther south
but is separated from probable Lower Boulder Clay by only 2.4 m
of loam, clay and sand. Moreover, an exposure of Lower Boulder
Clay [3116 4195] on the beach nearby is so high in the section that
little sand can be present. It seems likely that the southward rise of
the surface from Anchorsholme [320 420] to Norbreck is caused by
the thickening of the underlying sand, and marks the northern limit
of the Blackpool to Weeton ridge.

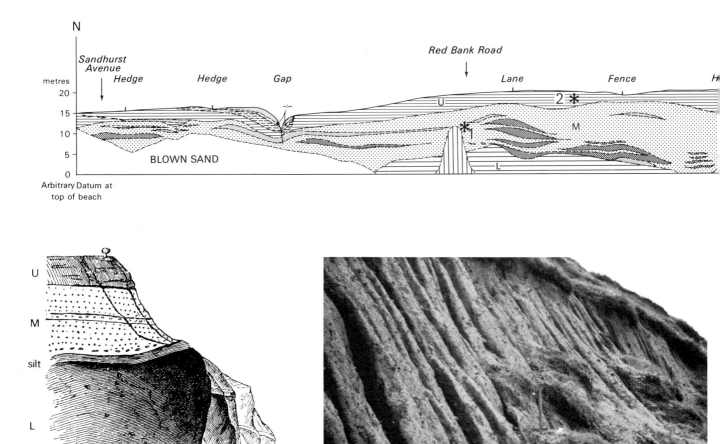

Note: the top of the section is not necessarily
the highest point along the coastline

Horizontal scale
Vertical exaggeration X 5

Figure 18   Cliff sections of Glacial deposits, north of the Boating Pond, Blackpool

Above is a hitherto unpublished section of the cliffs at Blackpool North Shore based on De Rance's field
notebook. Below are two of his published woodcuts, with two photographs of glacial deposits taken in 1968.
All sections are now obscured. The locations of illustrations 1–4 are shown by numbered asterisks on the
section. Illustration 2 shows badland weathering in Upper Boulder Clay. Illustration 3 shows the Upper
Boulder Clay overlying Middle Sand with lithified masses of sand

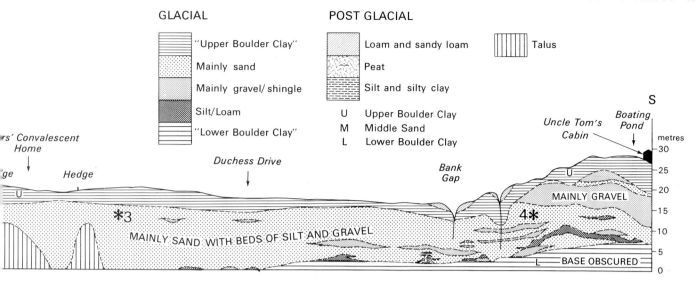

GLACIAL

	"Upper Boulder Clay"
	Mainly sand
	Mainly gravel/shingle
	Silt/Loam
	"Lower Boulder Clay"

POST GLACIAL

	Loam and sandy loam
	Peat
	Silt and silty clay
U	Upper Boulder Clay
M	Middle Sand
L	Lower Boulder Clay

Talus

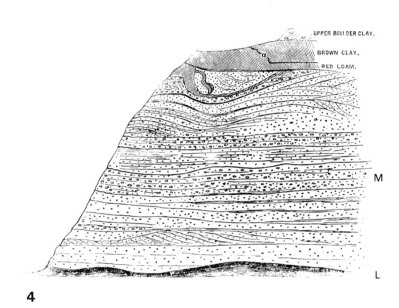

4

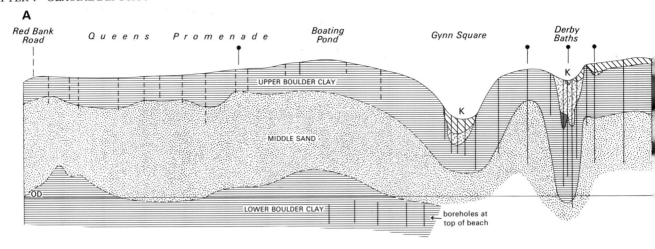

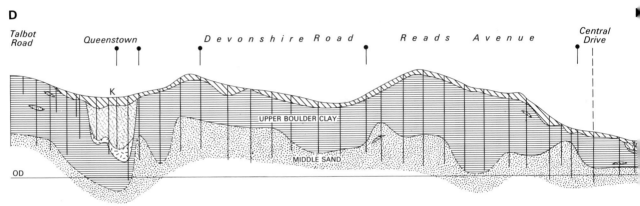

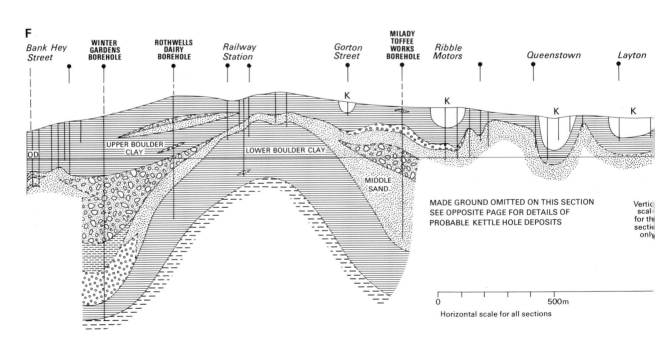

Figure 19 Horizontal sections in the Glacial deposits of central Blackpool.

Note the continuous mantle of Upper Boulder Clay, its downbuckling beneath probable kettle holes and the widespread but variable Middle Sands.

See also Figure 18 for more detail north of the Boating Pond

C

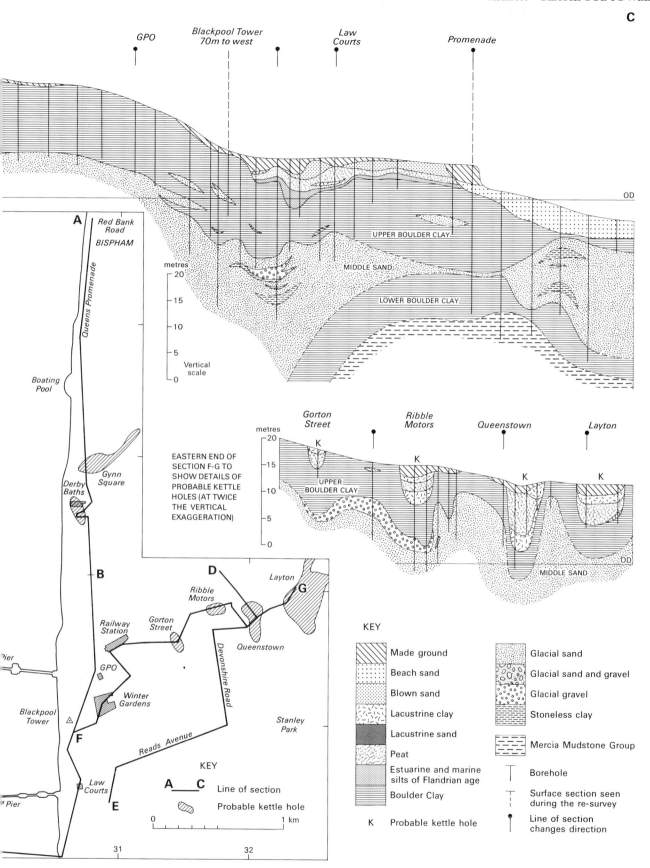

GPO

Blackpool Tower
70m to west

Law
Courts

Promenade

OD

UPPER BOULDER CLAY

MIDDLE SAND

LOWER BOULDER CLAY

metres
— 20
— 15
— 10
— 5
Vertical
scale
— 0

A
Red Bank
Road
BISPHAM

Queens Promenade

Boating
Pool

Gynn Square

Derby Baths

EASTERN END OF
SECTION F-G TO
SHOW DETAILS OF
PROBABLE KETTLE
HOLES (AT TWICE
THE VERTICAL
EXAGGERATION)

metres
20

Gorton
Street

Ribble
Motors

Queenstown

Layton

K

K

K

K

UPPER
BOULDER CLAY

15

10

5

0

OD

MIDDLE SAND

B

D
Layton

Ribble
Motors

Gorton
Street

G

Railway
Station

GPO

Winter
Gardens

Queenstown

Devonshire Road

Stanley
Park

Blackpool
Tower

F

Reads Avenue

Pier

Pier

Law
Courts

E

KEY

A _____ C   Line of section

Probable kettle hole

0   1 km

31   32

KEY

Made ground

Beach sand

Blown sand

Lacustrine clay

Lacustrine sand

Peat

Estuarine and marine
silts of Flandrian age

Boulder Clay

K   Probable kettle hole

Glacial sand

Glacial sand and gravel

Glacial gravel

Stoneless clay

Mercia Mudstone Group

Borehole

Surface section seen
during the re-survey

Line of section
changes direction

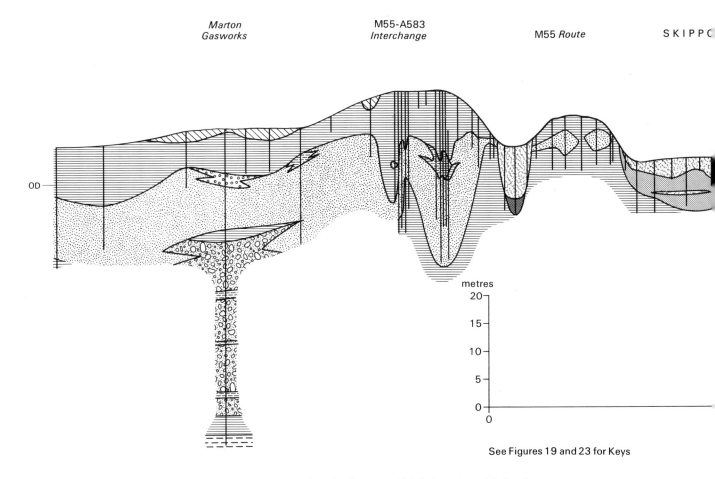

Figure 20 Longitudinal section based on borehole data in the superficial deposits, chiefly along the route of the M55 motorway

The cross-section (Figure 19) shows that many boreholes in Blackpool prove that the sequence in the cliffs continues beneath the town. The following log [3069 3605] is typical of the many recent site investigation boreholes hereabouts:

	Depth m
Boulder clay (Upper Boulder Clay)	to 19.0
Sand; silty, brown, on fine- to medium-grained sand	to 21.5
Sand; silty, brown, with thin bands of silty clay	to 23.7
Sand; fine- to medium-grained, denser downwards (unbottomed)	to 30.05

The principal lithology proved by the boreholes is sand with some gravel, and with local lenses of silt, clay and 'stony clay' which is probably till. The thickest sands generally occur where the complete drift sequence is thickest; for example, 48.16 m of sand in Winter Gardens Borehole [3090 3620] with an intercalation of 10.67 m of loamy clay, 43.6 m (unbottomed) at Milady Toffee Works [5155 3683] (Figure 19) and 44.2 m at Marton Gasworks [3449 336] (Figure 20). Locally, thick sands are present in comparatively thin drift sequences, as in a borehole [3007 3505] on the shore off Manchester Square, which proved 20.4 m of sand in an unbottomed drift section to 28.3 m.

Boreholes are more scattered away from the built-up area but Churchtown Borehole [3256 4056] is of interest in suggesting that the sand splits towards the edge of the deposit. Its log is: boulder clay 5.7 m; shelly sand 1.0 m; clay 3.2 m; shelly sand 4.3 m; boulder clay 8.3 m; sand 2.7 m; boulder clay 3.3 m. Another deep hole, ICI's B4 Borehole [3690 3870], proved 41.2 m of sand and gravel in a total drift thickness of 53.6 m.

Exposures are poor inland. North of Stanley Park, roadside sections formerly showed 9 m of sand with thin seams of silty clay, while cross-bedded sand with seams of pebbles was seen by De Rance beneath till in pits [3270 3499] in Great Marton. The best inland section at the time of the present resurvey was at Greaves Bros. sand-pit at Hardhorn [352 372], which exposed the highest 3 m of the sand and its till cover; the pit is now disused. Much of the working was in clean sand, but there were bands of till and laminae of silty sand in parts of the section. In one place, a steeply inclined and buckled layer of till occurred within the sand; in a second section, a till layer 0.05 to 0.13 m thick within the sand contained small lenses of sand, and overlay sand with layers of ochreous clay that were similar to those described by Taylor (1958) as cemented shear-planes.

Farther to the east, there are several exposures of sands on the flanks of the Skippool Channel. On the west side, there are extensive rabbit-warrens in sandy bluffs up to 6 m high [3703 3569]; scat-

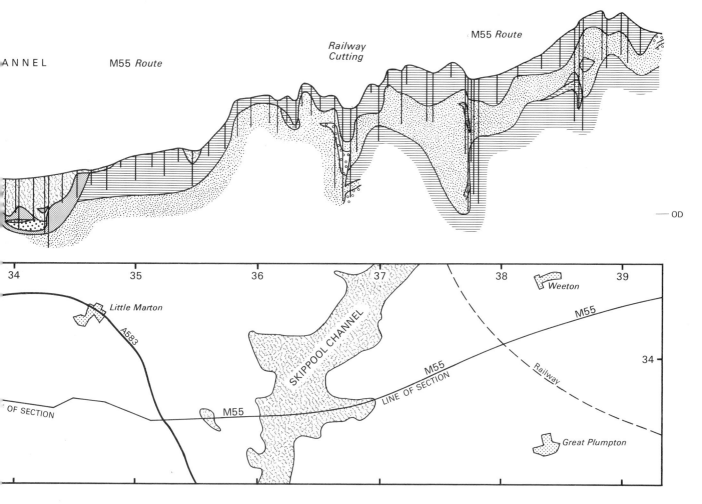

tered small pebbles resembling dreikanter are thrown out together with much fine-grained sand. Farther north, a sandy bluff [3705 3665] near Toderstaffe Hall is rich in cobbles. On the east side, the full thickness of sand, here about 12 m, crops out at Holme Wood [3755 3577] where fine- and medium-grained sand was formerly worked in a pit that is now grassed over.

Boreholes in the south-east of the district show variable thicknesses of Middle Sand. For instance, in B5 Borehole [3628 3871] gravels are 2.4 m thick yet only 600 m away, in B4 Borehole [3692 3870], gravels with some sands are 41 m thick. Farther south at Mereside Mushrooms [352 360], 26 m of gravels and sands were not bottomed. Farther east at Weeton Camp, many shallow boreholes prove 1 to 4.3 m of silt with bands of silty clay, locally passing into sand and gravel. Overgrown sand pits [3870 3413], 2 km farther south at Derby Hill, contain reddish brown, clayey sand in burrows.

In the extreme south-east, there are thick sands in two railway cuttings [3845 3360] near Weeton. There is no continuous exposure but fine- and medium-grained sand with some pebbles is thrown out in burrows on the sides of the cuttings, which are 5 to 9 m deep. The same sand was proved in the boreholes for the M55 motorway, about 1 km to the north (Figure 20).

Upper Boulder Clay

The Upper Boulder Clay occupies most of the surface outcrop of the glacial deposits. It is a brown, silty clay with abundant erratic pebbles and cobbles; no boulders were noted during the survey. The erratics are chiefly Lake District volcanic rocks, granites, greywackes and some Carboniferous sandstones. Fragments of marine shells are quite common. The till contains thin sand layers locally, particularly in its basal 1 m. Thicknesses are greatest in a belt running south-eastwards from North Pier, Blackpool to Mereside, within which they average about 11 m. Lesser thicknesses of around 3 to 5 m are usual to the north-east and south-west of this belt.

Data from many hundreds of boreholes in Blackpool and along the route of the M55 motorway reveal that the till is draped over the irregular upper surface of the underlying glacial sands. Thus, the irregularities of the sand–till interface are broadly duplicated by the present-day topography (Figures 19, 20). Among such irregularities are gentle topographical hollows floored by peat (see p.57), beneath which the till–sand contact also forms a hollow. Topographically, these resemble salt subsidence hollows, but most lie well outside the areas of possible salt subsidence. They probably formed by the melting of masses of ice within the glacial sands, with consequent collapse of the overlying till and the formation of hollows. Examples of such hollows occur at Queenstown Recreation Ground [320 367], Layton Highways Depot [324 382], Gynn Square [308 379] (all on Figure 19), and Pembroke Gardens Hotel [3068 3755]; a particularly large one [316 411] lies between Norbreck and Churchtown. Many of the depressions contain later deposits that range well up into the post-Glacial succession.

The south-eastern corner of the district near Great Plumpton is especially hummocky and, in places, the underlying Middle Sand crops out. However, the M55 section (Figure 20) reveals that even here the sand–till interface tends to duplicate each bump and hollow in the landscape.

DETAILS

Good coastal exposures were seen near Bispham during the present resurvey, but they have recently been hidden by landscaping. Scars in 4.3 m of till were locally fretted into 'badland' forms [3073 3964], the contact with the underlying sands being clearly visible. A discontinuous band of sand and pebbly sand up to 6 cm thick occurs some 0.70 to 0.85 m above the base of the till [3073 3958] (Figure 18). The cliffs northwards to Little Bispham are in till, which appears to be at least 6 m thick [309 406] west of Norbreck, but slips along the cliff face now obscure the section.

Inland, the basal part of the boulder clay was formerly well exposed for 30 m along a face at Greaves Bros. sand-pit, Hardhorn [352 372]. The till was 0.6 to 4.6 m thick with a markedly undulating base; in a small area, 0.25 m of sand lay in the lower part of the till.

Exposures of the till were also seen in two disused brick-pits. One, at Marton Moss Side [330 340], was in 6 m of reddish brown boulder clay containing shell fragments. The other, at Warbreck [320 381], was in 2.4 m of till which was formerly worked down to its total thickness of 6 m; a borehole here has proved the underlying sand. Many small exposures occur in disused marl-pits scattered over the fields. Two of the best are at Normoss, in 6.7 m of stony clay [3460 3748], and north-west of Brackinscal in 5.5 m of till (unbottomed) [3877 3773]. Other exposures are minor and temporary, generally showing red sandy and stony clay. The many borehole records are typified by the cross-sections in Figures 19 and 20.

Glacial Lake Deposits

Marton Mere is all that remains of an old glacial lake. It is now only 400 m long but 18th-century maps suggest that it was then much larger and two coracles, together with a metal anchor, have been found in fields that are now 1 km from the lake (Eyre, 1961). A terrace of fine-grained sand, 0.9 m of which are exposed, flanks the hollow in which the lake sits. It extends as a former beach deposit for about 1.5 km along the north side of the depression and an equivalent feature is cut into the till slopes south of the mere. The terrace flat lies at about 12 m above OD. This is higher than the two exits from the hollow, implying that these exits were blocked by ice when the terrace deposits were laid down. A second lake-level lies at 6 m above OD, and cuts into the frontal edge of the terrace on the north side of Marton Mere and into till on the south side.

Most of the deposits on the floor of the old lake are comparatively modern lacustrine alluvium. Beneath the surface, the alluvium probably ranges down into the late-Glacial. The deepest of several boreholes proved very soft, dark clay to an unbottomed depth of 9.8 m. Traces of peat were found in the clay between 3.7 and 7.6 m depth in this and nearby boreholes.

Interpretation of layered sequences

The interpretation of layered sequences of the type preserved in this area has varied over the years. Binney (1852) and De Rance (1877) seem to have considered that the Lower Boulder Clay was deposited from floating sea-ice; they postulated a succeeding marine transgression so that the Middle Sand was 'deposited around the coasts of an open but gradually deepening sea' that reached up to at least 450 m above OD; finally a recession of the sea and a renewal of cold conditions was thought to have been responsible for a further cold episode, with coastal ice producing the Upper Boulder Clay.

Over the years, this glaciomarine hypothesis was gradually abandoned, particularly once it was accepted that the marine shells within the drifts were no more than erratics, and as belief waned in a marine transgression of such dimensions. The two tills were then considered to represent two distinct advances of land-ice, separated by a recessional episode when diachronous sands were deposited widely over the Lancashire plain as moraines, outwash aprons and fills of pro-glacial lakes. The naming of the ridge extending eastwards from Blackpool as the 'Kirkham Moraine' (Gresswell, in Steel and Lawton, 1967) owes much to this thinking. Central to most such reconstructions was the concept of advancing and retreating active ice-fronts that were essentially water-tight. The main disputes centred upon whether the two till complexes were part of one composite glaciation or of two distinct glacial stages and, in the latter event, whether the Middle Sands were chronologically linked to the earlier or later glacial episode.

In turn, this interpretation came under attack when it was suggested (Carruthers, 1939) that multilayered sequences could form from the in-situ melting of a single ice-sheet, a view that has since been confirmed and elaborated by Boulton (in Price and Sugden, 1972). Accumulating geotechnical data has reinforced visual examination in distinguishing heavily compacted, over-consolidated, basal tills exemplified by the Lower Boulder Clay, from the much less compact tills typical of much of the Upper Boulder Clay. This difference was tentatively interpreted (Evans and Wilson, 1975) as a distinction between lodgement tills laid down at the base of an ice-sheet and those tills formed by in-situ melting of englacial and supraglacial debris of the same ice-sheet; the entire multilayered sequence was thus seen as the product of a single glacial incursion. More recently, this view has been supported by Longworth (in Johnson, 1985), though he also resurrected the idea that a shelf of sea-ice had been involved in the glacial history of the region, while Earp and Taylor (1986) have postulated that a similar succession in north Cheshire is the product of an ice-cored 'moraine'.

None of the ideas that have emerged in glaciology in recent years conclusively resolves all the problems inherent in the drift deposits of Area A. The compact Lower Boulder Clay is mainly a lodgement till, probably of Devensian age. All boreholes penetrating to solid in Area A prove this lower till, so it must be widespread. The less compact Upper Boulder Clay, also of Devensian age, is draped over the Middle Sands in a nearly continuous sheet. It seems likely that the Upper Boulder Clay was an ablation moraine which settled on the upper surface of the Middle Sands as the ice slowly melted.

The Middle Sands were formerly well seen in the Bispham cliffs to have cross-bedding consistently dipping to the south-east, indicating derivation from the north-west. The small-scale sedimentary structures show that the sands were deposited largely in water. Large blocks of ice were buried in the sands during deglaciation, and boreholes show that the Upper Boulder Clay sags beneath peat-filled kettle holes, where it was let down gradually as the ice melted. Though locally there are signs of disruption and collapse in the sands similar to those recorded by Longworth (*in* Johnson, 1985), the upper contact of the sands is commonly undisturbed and shows parallel layering of sand and clay. This suggests that the Upper Boulder Clay is unlikely to represent a readvance of the ice, for this would have involved shearing and widespread disruption of the sands.

The exact environment of deposition of the sands is the subject of speculation. The problem is to emplace these extensive deposits, which seem to be the products of deglaciation, without having to invoke a major readvance of the ice-sheets to deposit the overlying Upper Boulder Clay. One hypothesis envisages a mass of cold-base ice covering the drumlin terrain of Area B in the north and abutting against a stagnant mass of wet-base ice to the south (cf. Evans et al., 1968, p.245). Deposition of the sands in the early stages was probably on or around dead-ice, but may have continued on the land surface in the later stages. Following this explanation, the Upper Boulder Clay might then be expected to be mainly flow till and indeed some of it is. But most of it has characteristics unlike those of flow tills. For example, the Upper Boulder Clay tends to be consistent in thickness on both hill-tops and slopes, whereas flow tills are generally much thicker at the foot of a slope. Similarly, flow tills tend to fill hollows left by melting ice during deglaciation but many of the local hollows were left open at that time; only subsequently did they accumulate their peat deposits.

An alternative hypothesis seeks to account for the distribution of the sands by depositing them in water beneath the melting ice-sheet during deglaciation. But the sands are continuous over many square kilometres, so how could a cavity be produced beneath the ice of sufficient size to receive so much sand? What would support the ice-sheet roof above such a large cavity? These objections are met if the ice-sheet was floating on meltwater and the sand was gradually washed in beneath it. In the final stages of deglaciation, in this view, the water drained away and the ice-sheet melted to deposit the Upper Boulder Clay as an ablation till over most of the district.

The sands are derived generally from the north in the present district. A westward movement seems to be indicated in the sand occurrences around Preston, so the Blackpool sections may record a locally anomalous movement of sediment.

In Area A, there are relatively few boreholes but, in places, they show that there are deep hollows in rockhead. In the hollows, the Middle Sands are particularly thick (Figure 18) but the Lower Boulder Clay is still present beneath them. Certainly the two thickest sand sections, at the Winter Gardens and Marton Gasworks, are underlain by Lower Boulder Clay. The data is too sparse to show whether these are enclosed hollows or channels, but it seems likely that at least some of them originated as subglacial channels, which did not fully penetrate the Lower Boulder Clay and were eventually choked with sand.

In contrast to these doubts, the dating of the main part of the drift complex is now firmly established. One of the kettle holes in the surface till contains a coarse-detrital mud that has been dated variously at $11\,665 \pm 140\,\mathrm{BP}$ and $12\,200 \pm 160\,\mathrm{BP}$ (Hallam et al., 1973); the inference is that the buried ice began to melt a little before this date, and that the complex is late Devensian in age. Whether any of the unexposed lower part of the Lower Boulder Clay is appreciably older remains uncertain.

AREA B—THORNTON TO PREESALL

North of the area of multilayered drift, Area B is a small drumlin-field divided by the estuary of the Wyre (Figure 17). To the east of the river, the drumlins are best developed over the saltfield and around Staynall. They probably accumulated beneath an advancing ice-sheet that had a cold base. To the west, they are prominent south of Stanah, and several rise through the blanket of postglacial Marine and Estuarine Alluvium that occupies much of the ground north and west of the village; it remains uncertain how many buried drumlins there are hereabouts. It is difficult to delimit the drumlin-field accurately. The drumlins rise above a gently undulating sheet of till that occupies the interdrumlin tracts; the sheet merges eastwards imperceptibly into Area C. To the south, the low interfluves between Norbreck and Poulton-le-Fylde may be poorly developed drumlins smothered by flow till or they may be northward extensions of the Blackpool ridge. Inevitably, the limits taken for the area (Figure 17) are arbitrary.

Between Preesall and Hambleton, the larger drumlins trend at $150°$ to $170°$. On average, they are about 500 to 600 m long, about 200 m wide, and rise to about 20 m above OD. The greatest height attained is slightly over 30 m above OD on Preesall Hill [367 474], which is apparently a drumlin, though an atypical one. West of the Wyre, the drumlins maintain the same orientation and, where not modified by marine erosion, the same general dimensions. The drumlins rest on rockhead, and within the saltfield the major drumlins tend to occur where rockhead is deepest, as though some of the drumlins have 'roots', having been built-up over areas of glacial gouging. Even in such a tightly drilled area, however, it is not possible to reconstruct the rockhead surface sufficiently accurately to confirm this impression.

Boulder Clay and Glacial Sand

Surprisingly little is known about the lithologies that make up the drumlins. Most of the sections along the Wyre are in boulder clay. In the saltfield, there is one good exposure [3613 4665] of thick till within a drumlin, but most of the brine boreholes were drilled quickly through the drift and the drift samples do not generally reflect the drift sequences with accuracy, especially where caving of the sides of the boreholes has occurred. Some of the recent boreholes, such as Staynall and Coat Walls boreholes, passed through drumlins composed almost wholly of boulder clay. Their logs are too generalised to enable distinctions to be drawn between poorly consolidated tills in the Upper Boulder Clay and overconsolidated lodgement tills in the Lower Boulder

Clay, though extremely hard till is recorded in some logs for about 5 m above rockhead.

Boreholes are much more scattered outside the saltfield, but since they were generally drilled to provide specific constructional data, their records of the drift are more accurate. As in the saltfield, few holes have been drilled on drumlins; nevertheless, three CEGB holes between Stanah and Hambleton, sited on drumlins west of the Wyre, were mostly in boulder clay which increased spectacularly downwards in SPT (Standard Penetration Tests) values at a depth of about 4 m. This suggests that many of the drumlins are cored by compact till of Lower Boulder Clay type.

There are, however, several sections showing lenses of bedded sand within drumlins. The thickest sand underlies Preesall Hill where, although exposures are now poor, it may be up to 15 m thick near the hilltop. The most interesting section is along the Wyre estuary immediately west of the golf links at Knott End-on-Sea, where a longitudinal section through the western flank of a drumlin is exposed in the cliff. A compact till of Lower Boulder Clay type forms a low dome, rising parallel to the surface of the drumlin. It is overlain by about 5 m of false-bedded, shelly sand which is itself capped by poorly consolidated sandy till typical of the Upper Boulder Clay. The entire section is strongly reminiscent of the cliff-sections at Blackpool.

The interdrumlin tracts cover considerable areas of low-lying ground but there are practically no exposures. Ploughed fields are in sandy till with erratics up to about 15 cm across. Several boreholes within the saltfield record a few bands of sand and gravel within the till between the drumlins. The bands are usually barely 1 to 2 m thick, and are apparently distributed at random through the total drift thickness. Exceptionally, there are one or two records of about 6 m of sand, and one hole records 13 m. These records may well be unreliable, but appear to be on the flanks of drumlins.

There are no Glacial Lake Deposits and few glacial channels in the area. Small fans [348 415 and 377 438] at Little Thornton and Sower Carr, which have been mapped as Alluvial Fans (see p.50), may be late Glacial in age.

Details

Most of the drumlin tract is agricultural land, and red stony Upper Boulder Clay is everywhere turned up by the plough from beneath a thin surface layer of loam that becomes appreciably thicker between the drumlins. Many of the erratics have been collected around the field boundaries and farm buildings; there is a typical pile of erratics [3593 4748] just south of Curwen's Hill Farm, consisting of pebbles almost exclusively Cumbrian in provenance and up to 0.6 m across.

The freshest inland section [3614 4665] has resulted from subsidence near the Preesall salt-mine, which has exposed about 9 m of red-brown, stony, sandy till in the north face of a substantial drumlin; the section seems to be wholly in Upper Boulder Clay, though much of it is inaccessible. Up to 2 m of similar boulder clay are exposed in a small digging [3856 4534] just to the east, and the till has also been thrown out from the deeper ditches and from a few sewer and pipe-line trenches, notably one crossing the area from east of Skippool to Stanah Clough [3742 4000 to 3561 4300].

The most extensive sections are along the banks of the Wyre around Shard Bridge and for about 3 km downstream; here, the river winds its way between drumlins so that most of the sections are parallel to their long axes. The cliffs appear to expose at least 6 m

[3617 4218] and possibly 15 m [3729 4100] of Upper Boulder Clay. This is probably not the true thickness of the till since it is draped over a core of sand or Lower Boulder Clay rising beneath the axis of the drumlin. This is confirmed by several sections. For example, south of Staynall, the till/sand contact in an exposure of sand at the river-side [3600 4360] is clearly inclined south-westwards almost at the slope of the surface. Other minor exposures of sand [3644 4174 and 3656 4162] in a neighbouring drumlin appear to have a similar relationship to the overlying drape of till.

By far the best section across a drumlin is the one alongside the Wyre near Knott End-on-Sea [3462 4800 to 3483 4678]. Much of the section is obscured by slips along the cliff-face and by beach material. At its base, is an unbedded, highly compact, purple till of Lower Boulder Clay type. The included erratics are commonly quite small, about 25 mm across, but there are a few larger ones, principally tuff and granite from Cumbria, that range up to about 0.5 m. The till is overlain sharply by strongly false-bedded sand that is largely stoneless. Up to 4.5 m of sand is the maximum thickness seen at any one point, but the total thickness of sand in the section is difficult to determine because, although the contact with the basal till is sharply marked by a spring-line, the overlying till has slipped or flowed over the sand. Locally, 0.3 to 1.0 m of cemented gravel, rich in shell fragments and small Cumbrian erratics, lies at the contact between sand and basal till. The contact is sharp but at one point [3460 4725] a leaf of till wedges up into the gravel. The sand contains thin layers of purple clay and fine gravel, particularly near its top and these bring out minor contortions, most noticeably at the top of the section; there are also several microfaults. At one point, an old kettle hole contains 0.9 m of laminated clay which rests on sand and is apparently overlain by thin peat. The resemblance of this drumlin section to that recorded from the Blackpool cliffs, even to the cemented gravel at the base of the sands, raises the possibility that there may be drumlins within the tract of layered glacial sequences between Blackpool and Weeton which are wholly buried beneath the exceptionally thick Upper Boulder Clay.

A borehole [3468 4746] only 40 m from the cliff section through the drumlin shows how variable the sand layer can be, since it records only 2.4 m of sand between 6.1 and 8.5 m; the underlying 30.2 m of drift in the borehole is mostly till.

On the south-east side of Preesall Hill, a disused sand-pit [3688 4721] formerly exposed some 6 m of yellow-brown sand. The sand outcrop on the east side of the hill is quite narrow, with a strong spring-line below it that may indicate the position of the basal till. Near the sand-pit and an adjoining road cutting, however, the upper surface of the sand rises from the level of the alluvium at about 5.5 m above OD to almost 25 m above OD. near the summit of the hill, and the spring-line below the sand is arched in a similar fashion, suggesting that the sand forms the core of the drumlin.

A third example of a possible sand-cored drumlin lies at Burn Hall [3335 4445], west of the Wyre, where a low hillock rises about 9 m above the surrounding alluvial flat. An outcrop of sand circles much of the hillock; only on the eastern side is it not visible at the surface, and even here a strong spring-line probably marks its contact with basal till beneath a drape of Upper Boulder Clay. The sand is probably only 3 to 5 m thick at most. One small section [3320 4425] shows 5 cm of compact till lying at the base of the Upper Boulder Clay and resting on 0.5 m of fine gravel.

Another section showing sand beneath till is exposed at Limebrest Farm sand-pit [3470 4171], just south of Thornton. The pit lies on the western flank of a rise which is otherwise shrouded by till. The section shows about 1.5 m of upper till, dark and organic-rich in its upper part but red-brown below, resting on at least 3.7 m of sand. The top 0.6 m of the sand shows alternations of yellow and red, silty sand with scattered small stones. The banding picks out gentle contortions which maintain strict parallelism with the base of the till cover showing that, in this instance at least, the sand/clay interface is a sedimentary one undisturbed by sludging. The bulk of the sand is stoneless, but a little gravel appears in the west of the main face. About 400 m to the south, there is a small inlier of sand [3454 4128] in the western flank of another low rise, which probably also has a sand core.

Only till can be seen at surface on the other drumlins that rise through the alluvium. There is, however, a borehole [3244 4331] just beyond the exposed northern nose of one that proves a complex drift section, hard to interpret. Its log is as follows: clay, sand and gravel (?Marine Alluvium) to 2.3 m; sandy clay and gravel (?Upper Boulder Clay) to 4.4 m; alternations of sandy loam, gravel and sand (?Glacial Sand) to 19.8 m; clay to 21.3 m; gravel to 24.3 m; boulder clay and gravel (?Lower Boulder Clay) to 30.4 m (unbottomed). Even harder to interpret is the log of a borehole [3408 4370] apparently sunk on the crest of a low drumlin at Burn Naze, but where the surface is now covered by made ground. The log reads:

	Thickness m	Depth m
Boulder clay	12.19	12.19
Sandy marl	5.49	17.68
Sand and gravel	5.18	22.86
Sand, gravel and blue clay	1.22	24.08
Boulder clay	2.13	26.21
Gravel and clay	3.05	29.26
Stony marl	7.93	37.19
Sand and clay	0.61	37.80
Sandy loam	2.59	40.39
Sand	4.11	44.50
Running sand (unbottomed)	1.52	46.02

If the lithologies are accurately recorded, it is the most complex section of drift proved in the district. The section raises the possibilities that Quaternary deposits older than Devensian may be preserved at depth, or that some local sections are too complex to be fitted into the regional drift stratigraphy.

Despite these records, it is likely that sand-cored, topographical rises are the exceptions. Within the saltfield, surprisingly few boreholes are sited on drumlins and even fewer record detailed sections through the drift. Even so, it would be surprising if the presence of beds of sand had been ignored, since they usually affect the drilling, so it is probable that most of the local drumlins have till cores. Fortuitously, two BGS holes, drilled to prove the Triassic sequence, each began near a drumlin crest. Coat Walls Borehole [3551 4654] reached rockhead at 18.2 m, having been in stiff, brown and reddish brown, sandy boulder clay throughout, except for 0.9 m of sand with scattered pebbles at 5.70 m. Staynall Borehole [3562 4438] passed through 36.7 m of boulder clay before reaching 3.8 m of 'sand, boulders and clay' resting directly on rockhead.

Several boreholes north of Skippool have been drilled into drumlins. Three of them [3544 4207; 3557 4268 and 3599 4290] were in boulder clay for practically their entire depth. In the first hole, Standard Penetration Test (SPT) values for 0.3 m penetration, which are a measure of the degree of consolidation of the deposits, increased from 50–60 blows at 2.7 m to 120–160 blows below 6.4 m; in the second, a similar sudden increase took place between 3.5 m and 4.7 m. The third hole showed a rapid rise from 39 blows at 2.7 m to 78 at 4.3 m, and thence ranges between 77 and 90. East of the Wyre, a further borehole [3664 4322] near the crest of a drumlin south-west of Brick House Farm is in firm boulder clay, recorded as turning hard below 5.1 m; an SPT value of 145 blows is recorded at 8.95 m. These results show that many of the drumlins are cored by highly consolidated till, comparable lithologically with the Lower Boulder Clay and that this is overlain by much less compact Upper Boulder Clay lithologies. Where sand sections are present, they appear to separate these two types of till. Nowhere, however, is the borehole data sufficiently good to plot the interface between these lithologies.

AREA C — WHIN LANE END TO STALMINE

The spread of boulder clay extending from the Wyre northwards to the east of Hambleton and almost to Stalmine displays none of the layered sequences or drumlins that characterise areas A and B respectively. Instead, the till surface is gently undulating, the only distinctive features being numerous, closed hollows that range in scale from shallow, 10 m-wide depressions to large, flat-bottomed ones over 1 km across, though barely 3 m deep (Figure 21). During deglaciation, these hollows became shallow lakes; remnants of lacustrine fill and traces of exit channels with small associated fluvioglacial fans, still remain. Small though it is, the area provides a good example of deposition in a stagnant, dead-ice field.

Boulder Clay

Most of the few sections lie along the banks of the Wyre, especially where meanders are currently undercutting the cliffs. The best is in a bank about 12 m high [3843 4052] near Liscoe, which is apparently all in reddish, sandy till of Upper Boulder Clay type. Away from the river, surface sections are particularly poor. Most are in sandy or stony clay that is particularly sandy near the surface; in the many small depressions, some surface layers of stony sand, as at [3890 4350], approach mappable thicknesses. The contained clasts are small, but many of the farmyards [3957 4410; 3975 4168] are cobbled with Cumbrian erratics clearly gathered from the adjacent fields. The lithology and the general deposition of the till suggest that much of the surface material originated as supraglacial flow till. In a very few sections, notably around some of the larger hollows [3892 4322], the till is stiffer and includes thin seams of sand.

Only two boreholes in the area have reached rockhead. Hambleton Borehole [3820 4217] proved 29 m of boulder clay to rockhead. The other [3865 4146] is only about 800 m to the south-south-east, but its log is quite different: sandy clay to 10.4 m; soft sandstone with some gravel to 39.9 m; red marl with pebbles to 45.1 m; (undoubted) Mercia Mudstone. The record of 'sandstone' rather than sand suggests that this hole reached solid at 10.4 m, as does the record of a thin 'marl band' within sandstone at 22.9 m in another version of the log. However, there is no sandstone 29.5 m thick in this part of the succession, and the record is probably unreliable.

The other boreholes are all quite shallow. Several, for example at [3918 4274], have been drilled into the major depressions east and south-east of Hambleton Moss Side, and these proved silty, stony clay to their total depth of 6.45 m; small 'sand pockets' were recorded in one hole, and there were silty laminations at depth in another. The properties of the material drilled support the field observations which show that boulder clay everywhere lies near the surface within the depressions. Typical is a record [3884 4170] showing an abrupt increase in SPT values for 0.3 m penetration from 18 blows at 1.2 m (possibly lacustrine fill) to about 50 blows between 2.7 and 7.3 m; the latter figure is typical of much of the Upper Boulder Clay.

Only in one borehole [3964 4118], just to the east of the district, do SPT values suggest that basal till of Lower

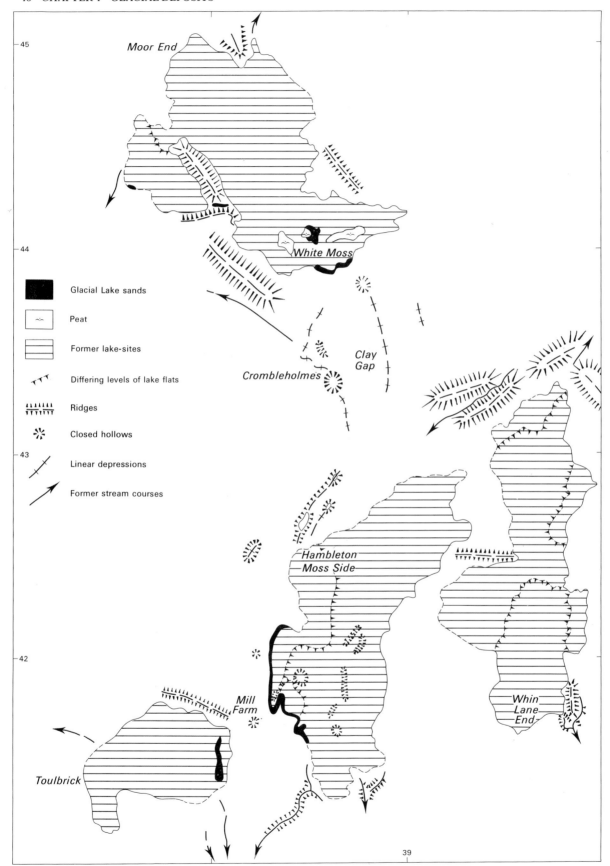

Figure 21 Sites of former glacial lakes in area C.

The extent of area C is shown on Figure 17

Boulder Clay type was reached. The log records an abrupt downward increase from 60 blows at 5.0 m to 93 blows at 5.8 m. Unfortunately, the hole terminated at this depth and it is not known whether similar values persisted to rockhead.

Glacial Lake and Fluvioglacial deposits

Although many of the larger, and some of the smaller, depressions in the till-sheet were once filled with standing water, the processes that initiated the larger hollows are not fully understood. It seems most likely that each of the depressions marks the site of a remnant block or blocks of ice that were particularly slow to melt. At a late stage in the melting, these remnant blocks probably stood up through more rapidly decaying ice, which deposited considerable thicknesses of flow-till around them. Consequently thinner drift was laid down beneath the blocks than around them, where the flow-tills accumulated more easily. On final melting, an inverted relief thus developed.

Several lines of evidence support this view. Thus, there is a suggestion of the presence of till-ridges aligned parallel to the margins of the depressions [3940 4321], (see also Figure 21); they may represent banks of flow-till laid down against the residual blocks of ice. Moreover, some of the depressions have several apparent exit channels each at a different level well above the later lacustrine flat; these are most easily understood if they originated as drainage off the surface of blocks of ice. Furthermore, the largest depressions are made up of several flats lying at slightly different levels; there is too little lacustrine sediment preserved for these flats to be depositional, and no appropriate outlets for them to be erosional; it is easier to envisage the sub-basins as forming from the final melt-out of discrete but contiguous blocks of ice.

The origin of the smaller, closed hollows is also problematical. Some probably result from irregular deposition, others as kettle holes in flow-till. It is, however, difficult to explain all the hollows in this way; this is particularly true of those that lie in the floors of larger depressions and appear to postdate the drainage of the shallow lakes that temporarily occupied them. Even harder to understand are some hollows which are now commonly flooded, but were apparently dry when the Ordnance Survey base-map was constructed. The inference is that the cause of the subsidence is contemporary, and it it probably due to the solution of irregular remnants of rock-salt in the underlying Singleton Mudstones.

Despite the considerable size of the larger hollows, the lacustrine deposits within them are rarely of mappable thickness. Boulder clay has generally been dug from ditches that reach to 1 m below the surface, though thin spreads of clayey loam, containing small stones and much organic residue from the formerly extensive peat, are common. The most significant deposit in the central part of a depression is a patch of stony sand [3852 4405] about 1 m thick; it covers an area of about 100 m × 100 m on White Moss near Hale Nook, and continues as a thin skin for another 400 m to the east. Boulder clay and blue clay were dug from a trench immediately to the south, and stoneless clay overlies boulder clay [3804 4448] farther to the north-west in the same depression.

Rather more common are narrow strips of silty sand or gravelly sand bordering some of the depressions. Some lie a little above the lacustrine flat, and some are associated with a slightly higher, terrace-like notch. The sands have been traced for up to 500 m [3831 4185], though their outcrops are only 10 to 20 m wide. Their likely origin is as lake beaches, but they may have been laid down in drainage systems that skirted the melting ice-blocks before the lakes formed, for some of them seem to be associated with drainage channels that lead through the rims of the depressions (Figure 21) but lie above the likely lake levels.

There are many such drainage channels leading away from the sites of the glacial lakes, and some of the streams that once occupied the channels — not necessarily the largest — laid down small fans of loam where they debouched on to lower ground. The best examples [3810 4100] lie east of Moors Farm; two small deltas of loam have accumulated where minor channels from one lake enter a small stream that drains south-westwards towards the Wyre from the site of another lake. The fans have been mapped as Fluvioglacial Deposits, though the drainage regime might best be considered as fluvioglacial. Similar fans, built up by meltwater originating in this tract, lie around Hambleton and Sower Carr (p.50).

CHAPTER 5

Post-Glacial deposits

The post-Glacial sediments are generally quite distinct from the Devensian deposits and are mainly marine and estuarine clays, silts, sands and gravel laid down in Flandrian times. Initially sea level was some tens of metres lower than at present and the water gradually rose as the ice sheets melted. Beneath Fleetwood thick marine alluvium extends down to 22 m below OD and at Blackpool South Shore to 16 m below OD. In the latter area peat occurs beneath the marine alluvium, indicating land engulfed by a rising sea. Gradually the sea lapped against the comparatively high ground of the Blackpool-Weeton ridge and the drumlin field.

It is not everywhere possible to separate the Glacial and post-Glacial deposits. This is particularly the case within the Skippool Channel and in the inland kettle holes and drainage channels, where deposition may well have begun before the late-Devensian ice completely melted. In most of these locations the surface sediments are Flandrian in age, and it is convenient, though perhaps strictly incorrect, to deal with the associated buried deposits in this chapter. Spreads of peat and river alluvium postdate the marine incursion, and contemporary or near-contemporary storm beaches, sand dunes and salt marshes complete the Quaternary history of the district.

Alluvial Fans

An extensive alluvial fan is preserved at Sower Carr [377 438] and abandoned stream channels meander over its surface. The deposit is mostly clay, with some silty clay containing a few small pebbles. It is quite thin, for several drainage ditches have reached the underlying till at about 1 m. Even so, the fan is too large to have been laid down by contemporary drainage. The feeding channel leads from an area to the east [384 436] near Crombleholmes, characterised by dead-ice drainage features. The channel was probably operative when the ice was melting; if this was indeed the case, the fan would be better classified as fluvioglacial. This view of the origin of the fan is supported by the way in which the Estuarine Alluvium seems to cut into the front of the fan along its southern side, though the two deposits seem to merge a little farther north.

A similar, but smaller, fan a little to the south [372 432] appears to be contemporaneous with the Estuarine Alluvium but is linked to a terrace of silt standing just above the alluvial flat. Again, the feeding channel seems to have carried melt-water.

West of the Wyre, a small fan near Little Thornton is divided [3485 4150 and 3460 4142] by a minor stream which grades to the Estuarine Alluvium. The sediment was almost certainly fluvioglacial and a late-Devensian age can be postulated.

Older Marine and Estuarine Alluvium and Older Storm Beaches

Much of the north of the district is covered by an almost flat spread of Marine and Estuarine Alluvium analogous to, and presumably contemporaneous with, the Downholland Silt of districts to the south (Wray and Cope, 1948). There are three main spreads. One lies inland from South Shore, Blackpool and is covered by later Blown Sand and Peat; the second extends northwards from Little Bispham and Thornton to Fleetwood; the third lies to the east of Knott End-on-Sea. Smaller tracts spread southwards along the Skippool Channel and eastwards from the Wyre along several of the intra-drumlin hollows.

South Shore, Blackpool

The extensive triangular tract of flat ground that extends southwards from Central Pier continues southwards to Lytham St Annes and the Ribble estuary. It is underlain at depth by Marine Alluvium of Flandrian age. The deposit is up to 15.75 m thick, but is everywhere concealed beneath Peat and Blown Sand. Its depth and lithology are, however, proved by several hundred sewerage and site-investigation boreholes. Contours on the base of the Marine Alluvium are given in Figure 22, and the more representative of the boreholes are shown in Figure 23.

The Alluvium rests on a surface cut by the sea across the low-lying parts of the glacial drift. The shape of the surface is best known between the South and Central piers, where many boreholes have entered the underlying boulder clay (Figure 22). This undulates from 4 m above to 4 m below OD, the Marine Alluvium filling hollows in the surface. Southwards from South Pier, boreholes are more scattered but show that the surface slopes south-westwards, the deepest proving being at 16.1 m below OD [3042 3324]. Just south of the district, there is a deep embayment in the surface of the boulder clay, the top of which is at 11m below OD on the northern perimeter of the airfield [3235 3287].

Patches of peat resting on, or very close above, this surface have been drilled on South Shore. They are 0.30 to 0.75 m thick and either rest directly on till or are separated from it by up to 0.90 m of silt or clay. The level of the peat relative to OD varies with the depth of the underlying erosional surface, down to levels as low as 16 m below OD and, because it is so patchy, it is not clear whether it is a single bed. It is noteworthy, however, that thin peat has been noted at the base of the alluvium elsewhere around the Lancashire coast; near Heysham, such an occurrence has yielded a date of 9270 years BP (Shotton et al., 1970).

The main body of Marine Alluvium is a silty sand or silt passing upwards into silty clay. Marine shells have been recorded from it in several boreholes, particularly those on Lytham Road. The deposit thins inland from a maximum of 15.75 m near the coast to a feather-edge in the east. Locally, under Lytham Road, just south of the district, the silts seem to pass laterally into 5.5 m of gravels, possibly an old storm beach deposit. These gravels appear to run in a belt north-north-westwards towards the Pleasure Beach [306 333], where up to 8 m of gravels have been proved in several boreholes (Figure 23, Section A–B).

Palaeontological studies have been carried out on samples of Marine Alluvium from a borehole [3364 2799] at Lytham Hospital and, though the site is about 5 km south of the district, the results are probably applicable to the Blackpool deposits. The hole proved Marine Alluvium to a depth of 19.70 below ground-level. The basal 0.80 m was a silty sand, from which Miss D Gregory identified foraminifera indicative of a cold sea, less than 20 m deep. This bed was overlain by 0.50 m of peaty clay, probably equivalent to the peat beneath South Shore. Mr M J Reynolds recovered a sparse pollen flora from it, in which tree pollen is represented by *Pinus, Betula* and *Corylus*, and grass is also present, together with rare herbaceous pollen. The flora, although sparse, is indicative of the cool climatic conditions of Zone I of the Flandrian. From a sandy silt a little higher, between depths of 16.10 and 17.95 m, Miss Gregory obtained another shallow-water fauna that points to some climatic amelioration. The cold-water species *Elphidium clavatum* is joined by such southern species as *Ammonia batavus, Bulimina gibba/marginata, Elphidium williamsoni, Nonion depressulum* and *Protelphidium anglicum*; the ostracods *Leptocythere* spp. are also comparatively abundant. Still higher, at 15.00 to 15.45 m, an association of common *E. williamsoni, P. anglicum* and *Trochammina inflata* in silty and sandy clay indicates fluctuating salinities and possible near emergence, in conditions very like modern estuarine mud-flats. Finally, the fauna from 8.60 to 11.95 m shows a return to rather colder conditions.

Thornton to Fleetwood

A nearly flat, clay-covered plain of Marine Alluvium, its surface lying at about 5 m above OD, extends northwards from Thornton to the coastal sand dunes at Fleetwood, and is broken only by a few low drumlins that rise through it. In the south, the Marine Alluvium extends as narrow tongues along minor streams draining off the Blackpool–Weeton ridge and merges imperceptibly into river alluvium. Nowhere are there any marginal storm-beaches similar to those found round Preesall, and the Alluvium seems to lap against a gently rising slope of till.

Surprisingly few boreholes in this tract have reached the underlying glacial deposits, and some of those that do, particularly along the west bank of the Wyre, appear either to have passed through considerable thicknesses of Made Ground (unrecognised as such in drillers' logs) or into very local channel-fills deposited by the river. There is, thus, too little information to determine the form of the sub-Alluvium surface, which seems to be irregular. In places, glacial deposits are quite near the surface. Indeed, an old borehole at Burn Naze [3408 4370], on a site now heavily modified, began at about 8 m above OD in boulder clay; surface contours, surveyed before the ground had been disturbed, suggest that the borehole was drilled into the top of a drumlin lying just above the Alluvium. Other boreholes, [323 456], passed into boulder clay at 2 and 4 m below OD; SPT values of the till were typical of the Lower Boulder Clay, and the

holes probably passed into the core of a truncated drumlin that is now wholly concealed by Alluvium. Other such drumlins may well remain undetected. A thin peat-bed is exposed from time to time along the foreshore [311 435] between Bispham and Cleveleys. Its position in the sequence is uncertain, but it may equate with similar patches proved on South Shore, Blackpool (see p.50) and lie at or near, the base of the Post-Glacial sequence.

In general, however, the glacial deposits lie well beneath the surface. The thickest records of Alluvium are in B8 Borehole [3225 4529] and in several offshore boreholes [309 423] just to the west of Anchorsholme, where it continues downwards to 22 m below OD. Northwards to Fleetwood, there is nothing to show whether these levels continue, other than several boreholes which were still in Alluvium at 12 m below OD. Around Wyre Dock, levels of about 14 m below OD are recorded, but farther north there seems to be a low ridge of glacial deposits, for a borehole [3411 4798] on the west bank of the Wyre enters boulder clay at about 6 m below OD, and one [3378 4844] at the North Euston Hotel appears to have met boulder clay at 8 m below OD.

The most intriguing records are from a group of boreholes near the coast at Cleveleys. They appear to demonstrate an abrupt step in the surface on which the Alluvium rests (see Figure 24). To the south, the Marine Alluvium rests on Boulder Clay at 1 m above to 2 m below OD [e.g. 3162 4287 and 3178 4284]; to the north, the contact falls to 18 m below OD [4139 4306]. It seems possible that this buried step continues eastwards across the peninsula marking a still-stand of the generally rising, early Flandrian sea. Along the northern slope of the Blackpool-Weeton ridge, the streams that drain northwards to the main alluvial flat may well have been graded to this lower sea level. Boreholes [e.g. 3279 4133] within the valley leading northwards from Churchtown, prove soft clays, silts and peats to at least 12 m below OD, though there is the added complication here of possible salt subsidence; another borehole [3427 4084], in the larger valley at Breedy Butts, appears to have been still in alluvial sands when it terminated at 5 m below OD.

There is some evidence, though it is inconclusive, that there is a significant change in the nature of the Marine Alluvium as this buried coastline is crossed. To the south of it, the deposit is almost wholly of grey, silty clays, weathering brown and containing thin silt laminae; the clays locally pass down into grey, clayey silts. This deposit lies at the surface over most of the peninsula and appears to have been laid down during the renewed marine transgression which trimmed the tops off some of the drumlins and overstepped the earlier coastline. North of this coastline, though not necessarily continuing to it, a suite of sands and gravels lies beneath the surface clays. The underlying till is overlain by fine- to medium-grained sand with much gravel, that passes upwards into grey sand with only a little fine to medium gravel. Records of the depth of the change from mainly gravel to mainly sand do not give a consistent picture and there appears to be no regular contact between the two. In B8 Borehole, for example, sandy gravel is recorded from 22 m below OD to 11 m below OD, and is overlain by 'clay and gravel' that extends up to 3 m below OD. At the Fylde Computer Centre, the gravel was not bottomed; its top lay at about 4m below OD, and the sand continued upwards to 0.6

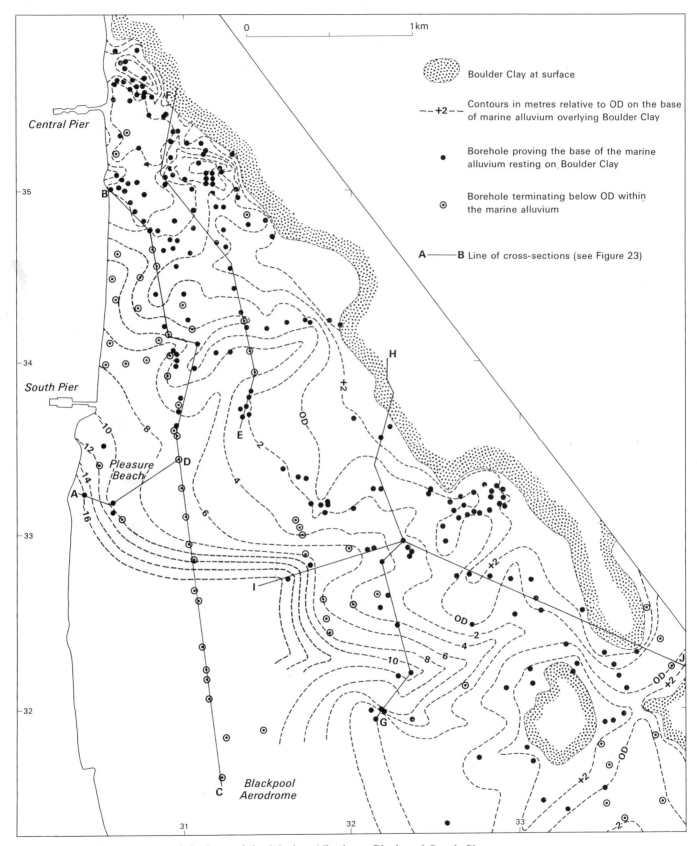

Figure 22   Contour map of the base of the Marine Alluvium, Blackpool South Shore

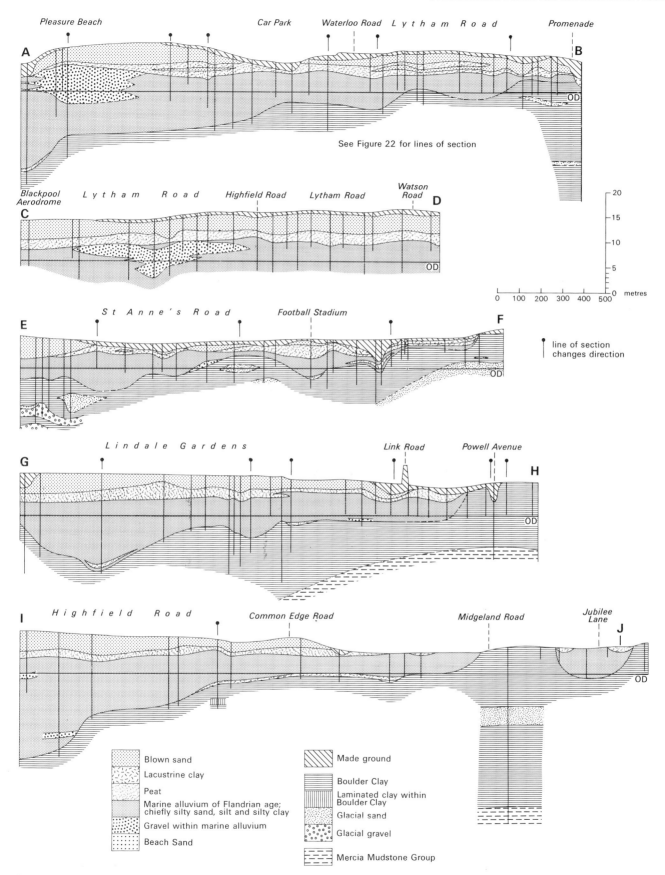

Figure 23   Sections across South Shore in post-Glacial deposits. See Figure 22 for lines of section

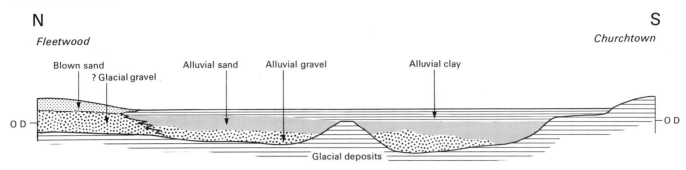

Figure 24   Generalised distribution of post-Glacial deposits between Fleetwood and Churchtown (not to scale)

to 1.2 m above OD. A little farther north [320 468], the surface clays and silts are underlain by sands at 2 to 3 m above OD, and gravels, with a few clay lenses, are then recorded to at least 8.5 m below OD. Near the Memorial Park, Fleetwood, the surface clays have been proved [329 472] to about 3 m above OD, and rest on sand with some gravelly bands. Gravel is also recorded in boreholes alongside the Wyre, particularly in one borehole [3411 4844] where its base was at 8 m below OD, with a further thin gravel at 14 m below OD: these latter provings may, however, be distinct from the main sheet to the south.

The extent of this sheet and its broadly horizontal disposition suggest that it is a marine deposit rather than a glacial one. The deposit is characterised by abundant marine shells, described as 'complete oyster-like shells' in the gravels. Whether the sheet built out northwards as a sandy tract rather like modern Dungeness, or whether it formed behind a spit that underlay much of Fleetwood and the west coast, remains obscure. If there was such a spit from Rossall Point to Anchorsholme, it has been subsequently destroyed, for along much of this stretch the surface clays continue westwards to the sea-defences, and a peat presumably within the clays is sporadically exposed beneath the beach-sands, [e.g. 3123 4371]. Low mounds of shingle [3131 4624 and 3137 4600] around Larkholme Farm may be remnants of such a bar, but it is not certain that they are natural deposits.

The general relationships of the deposits in the Marine Alluvium are shown in Figure 24. The implications of the latest changes in sea-level are discussed below (p.55).

East of Knott End-on-Sea

An almost flat tract of Marine and Estuarine Alluvium extends eastwards from the mouth of the Wyre and is linked to a smaller spread along the Wyre by a west-trending hollow crossing Hall Gate Lane just south of Preesall Park. The few boreholes that have bottomed the Alluvium are generally poorly recorded and unreliable. Modern records at Knott End [3530 4829], establish the deposit to be at least 6 m thick and to extend downwards to at least 1.5 m below OD. The deepest hole [3463 4835] may have reached the underlying glacial deposits at about 3.5 m below OD, where it entered clay much firmer than the near-surface clay, although the former may still be part of the Marine Alluvium. Farther east, the marine erosion surface at the base of the Alluvium presumably slopes gently northwards, and is probably at about 11 m below OD beneath Pilling Moss [395 460] and at about 18 m below OD near the coast at Fluke Hall [381 502].

At the surface, soft grey silty clay is dominant, though surface layers are commonly sandy near the coast where some of the sand may have been blown inland in the very recent past. Boreholes record shells and roots throughout the deposit but there is nothing to suggest the presence of substantial and laterally continuous beds of peat, although peaty layers were encountered in some boreholes. It proved impossible to confirm the presence of the peat noted by De Rance on the primary survey map (1874) cropping out on the beach northeast of Knott End beneath beach sand; it is presumably a local development within, or at the base of, the alluvial sequence.

There are very minor surface undulations on the coastal plain. Several slight rises, trending roughly east–west, are probably formed by seams of sand and silt within the clay. Notable examples run through Captain's Old Houses [3916 4870] and along Head Dyke Lane [3780 4715]. Here, up to 5.75 m of compact sand underlie loose Blown Sand at the surface and the two deposits cannot be separated. There is, however, nothing to suggest that there are laterally continuous beds of sand and gravel at depth, as there are south of Fleetwood.

South-eastwards from Knott End, several ridges of gravel are closely aligned parallel to the inner edge of the Older Marine and Estuarine Alluvium. There seems no reason to doubt the view that they are remnants of marine beaches (De Rance, 1875). Other ridges to the north appear to mark the coast as it retreated northwards, a retreat punctuated by temporary inundations of the nearly flat, coastal plain. These Older Storm-Beaches are shown in Figure 25. In the extreme east, several gravel ridges are more than 3 km long and extend eastwards into the Garstang district. East of Greengate Farm [393 453], three separate ridges run through the peat of Stalmine Moss. Gravelly sand, with clasts up to 12 cm across, has been exposed in a ditch [3944 4516] in the southernmost of these. The middle ridge is laterally the most extensive; it is 50 to 70 m wide and extends from a low bluff of till for about 860 m eastwards before the gravel passes beneath peat. The gravel can be traced still farther east because differential shrinkage of the peat over the gravel allows the surface ridge to persist as a topographical feature. The northernmost of the three ridges is a low gravelly rise [3935 4543] to the north of the farm.

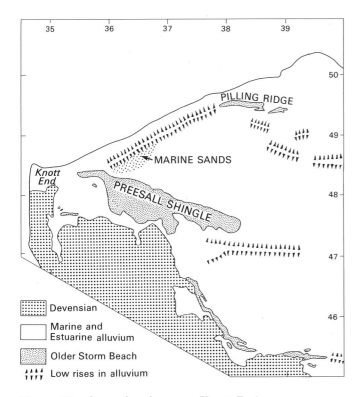

Figure 25 Storm beaches near Knott End

About 300 m to the west, a low rise in the surface of the Alluvium marks another buried ridge of gravel which leads to a belt of gravel outcrops about 80 m east of Old Tom's Lane [3833 4553]. The gravel continues north-westwards, amalgamating with another subparallel ridge to the south where the gravel can be seen in a ditch [3850 4546] along Crook Dale Lane. The clasts in the gravel are well rounded and mostly about 10 cm in diameter through ranging up to 60 cm. The combined ridges are 50 to 70 m wide and up to 3.5 m high, and there is another low rise of gravel close to the south-west. There is a break in the main gravel ridge [3802 4551] just north of Moss House but the gravel then continues, giving rise to an even sharper topographical feature and forming a bar across the Ivy Cottages Channel (see below) before terminating [3732 4662] about 300 m north-north-west of Burned House Lane. There have been several small gravel diggings along this stretch [3772 4606 and 3748 4642], and the clasts are up to 20 cm across. Farther to the north-west, apart from one small inlier of gravel, there is no sign of the deposit until south of Hackensall Road Crossing where there are four parallel ridges of gravel. The most continuous [3520 4720] is almost 350 m long. All these ridges are interpreted as marking the approximate limit of the Flandrian transgression, though the Alluvium generally extends 100 to 200 m 'landward' from the storm beaches and spits, and the sea may have coursed freely through the Ivy Cottages Channel before this route was blocked by gravel.

Another, much larger, sand and gravel deposit, probably slightly younger, extends for about 3 km south-eastwards from Knott End in an outcrop 3 km long and up to 600 m wide, though it rises only about 2 m above the Alluvium. It has been called the Preesall Shingle (De Rance, 1875), and

has been worked extensively in several pits [3613 4825, 3650 4810 and 3760 4775] north-east of Sandy Lane. Low subsidiary rises along its eastern margin [3781 4745] mark the position of discrete ridges that amalgamate westwards into a single spread, though even here the dentate inner margin of the outcrop suggests that many ridges are represented. At the time of survey, the disused pits were flooded and being filled. There were no good sections, the best [3758 4779] showing 1.5 m of gravelly sand with Cumbrian, and possibly Scottish, erratics generally 5 cm, and up to 15 cm, across. There was little sorting, and ovoidal clasts were uncommon. De Rance (1875) records the presence of several molluscs including *Cardium, Purpura, Natica* and *Turritella*. The maximum proved thickness of the gravel is 6.7 m in a borehole [3682 4774] in Sandy Lane, Knott End, where it extends down to 0.6 m below OD. Alluvial clays lap against the gravel around the margin of the ridges and one borehole [3593 4781] near Parrox Hall entered sand and gravel beneath 2.7 m of clay. It seems likely, however, that the main gravel wedges out abruptly both to the north and south of its outcrop.

East of this tract, a low rise of silt and sand along the line of Head Dyke Lane may overlie a buried continuation of the Preesall Shingle but there are no provings at depth. Other minor sandy ridges, generally trending just south of east, run through Smithson's Farm [3812 4845], Chestnut House [3922 4900], and Townson Hill [3958 4854] just east of the present district; there may be buried beaches or sand-bars splaying outwards from Knott End-on-Sea, but if so they lie beneath younger sediments.

Much nearer the coast, Pilling Ridge [3790 4954 to 3920 4945] is another, and still younger, ridge along which there have been several small diggings in shingly sand, with stones mostly less than 3 cm across but ranging up to 8 cm. East of Cocker's Dyke Houses, the ridge swings slightly, parallel with Pilling Lane but there are no provings at depth. At the western end of Pilling Lane, however, boreholes [3606 4871] have passed through a thin skin of Blown Sand into compact sand that was not penetrated at 1 m below OD. It seems possible that marine sand fills the hollows between the outcrop of the Preesall Shingle and that of the gravel along Pilling Ridge.

The significance of these Older Storm Beaches and of the Marine Alluvium, in terms of Flandrian sea-levels, is far from certain. The inner edge of the alluvium is at about 5.5 m above OD, a level that agrees closely with that of the postulated Hillhouse Shoreline (Gresswell, 1957). However, these strand-lines are likely to be controlled by high-water marks, and in this area current high tides rise to about 4.5 m above OD, only the coastal defences preventing repeated inundations. There is thus little evidence of a significant fall in sea-level since the Alluvium was laid down, and it is more likely that the coastal plain has been built outwards by marine aggradation.

Wyreside

Several tongues of Older Estuarine Alluvium extend eastwards from the Wyre. They follow hollows in the surface of the till that were probably initiated by melt-waters, and it is, therefore, likely that the sediments at depth within the hollows are appreciably older than the surface deposits. The most notable is one that almost links the Wyre near Barnaby's Sands to the spread at Preesall Moss Side, and is here referred to as the Ivy Cottages Channel. It is, indeed, possible that the break in the channel at Church Bridge [371 460] is an artificial embankment, and that the area around Preesall to the north was once an island. Clayey silt and silty clay floor the channel and, where it debouches onto the Wyre alluvium, ditches show seams of sand and silty sand [3581 4580 and 3563 4613] with a band of peat, up to 6 cm thick, close beneath the surface. It seems likely that high tides occasionally coursed through the channel well after the main fill was deposited, for there is a pronounced step in the alluvium [356 456] just north of Corcas Lane which seems to be erosional. The effects of sea-protection works, salt subsidences and agriculture, however, all combine to make any interpretation of this area speculative.

A narrow channel links this tract to another inlet of Older Estuarine Alluvium around Hambleton which extends inland to about 4 to 5 m above OD. In the narrowest part of the channel [3608 4408], there is probably only a thin deposit. Even where the channel is wider, a borehole [3695 4324] south-east of Brick House proves only 1.5 m of silty clay and silt, underlain by 0.46 m of peat with its base at 2 m above OD. It is not practicable to separate clay from silt in this terrain. Nearby houses [3722 4275] and others in Hambleton [3727 4226] have their foundations in stoneless, brown, sandy silt and sand, some of which may be glacial in origin, but all the deposits are retained as Older Alluvium.

In the extreme east, a delta-like flat [399 409] east of Liscoe Farm is at a level near to that of the Older Estuarine Alluvium and is categorised with it. Immediately east of the district, it almost links with the lowest terrace of the Wyre, while the saltings immediately south of it, though cut by minor steps that probably stem from tidal surges, seem to pass into the floodplain of the Wyre.

The Skippool Channel is a sinuous tract, about 400 to 500 m across, that links the Wyre just north of Skippool to the Ribble at Lytham, south of the present district. At surface, there are exposures of almost stoneless clay [3547 4025] and blue, estuarine alluvium [3555 4017] but these deposits pass southwards under the peat which floors most of the channel. The channel is crossed by the M55 motorway [363 336] near Moss House Farm, where boreholes suggest that its likely depth is 5 m below OD, while the bottom of a feeding channel west of Skippool is more than 5 m below OD. It seems likely that the Wyre has occasionally made use of the main channel in the past, probably at times of high tide and northerly winds. This is not to say, however, that the channel was formed by that river. It cuts through the Blackpool–Weeton ridge to a depth of at least 20 m, and near Weeton to nearer 40 m. It is hard to see how the Wyre, which in this part of its course certainly dates back only to late-Devensian times, could have carved this impressive channel, or abandoned it once it became operative. One interpretation of such drainage anomalies, much invoked in the past, is that they represent overflows from glacial lakes, but in this instance there is no evidence of such a lake. It seems more likely that the channel was cut by either supraglacial or subglacial meltwater flowing southwards on, or below, the surface of the ice, possibly following the course now taken by the lower Wyre, and gradually cutting downwards to the lodgement till. Whether any of the 'Middle Sands' had been laid down when the channel was operative is uncertain, but if the 'Skippool River' eroded and redistributed some of the sands, then the apparent lack of continuity of the glacial sands to the west and east of the channel (p.37) is easily understood. Similarly, the relationship of sand and till along the flanks of the Channel may be strictly local to this tract.

The deposits within the Skippool Channel have been proved in boreholes close to its northern and southern ends. In the north, near Little Singleton, five boreholes near the middle of the channel, proved at least 9.05 m of channel fill. Generally, silty clay up to 5 m thick overlies sand and gravel. In one borehole [3669 3904], peat, split by a clayey sand parting, lay between 5.0 and 5.9 m in depth. In the south, a series of boreholes drilled for the M55 motorway shows that at least 6 m of channel deposits lie beneath a surface blanket of peat. They consist dominantly of silts with a thin median peat in three of the boreholes. A little sand and gravel is also recorded within the sequence (Figure 20).

Peat

At the end of the Devensian, when temperatures rose and the ice had melted, peat began to accumulate in hollows on the surface of the till, and probably in the sinuous valleys around Marton Mere and in the Skippool Channel; deposition at some of these sites may have continued almost to the present day. Nearer the coast, a patchy cover of vegetation seems to have been established before being swamped by the Flandrian sea; it is preserved as the basal Flandrian peat (p.50). Impersistent, thin, peaty bands are common throughout the Marine Alluvium. Whether they are indicative of systematic minor fluctuations in sea-level or whether they are fortuitously preserved, local accumulations on an irregularly aggrading, coastal flat is uncertain. Finally, extensive spreads of peat, lying inland from South Shore, Blackpool, and east of Preesall, covered the surface of the Marine Alluvium but over the last century much of it has been removed by cutting.

Blackpool, South Shore

Drilling indicates that peat underlies the triangular coastal strip, up to 2 km wide in the south, that extends southeastwards from Central Pier. The area is contiguous with the main part of Marton Moss in the district to the south, where De Rance (1877, p.79) recorded trunks of oak, alder, yew, ash and fir. In the Blackpool district, the peat is about 2 m thick but in places reaches 3 m. A lens of Blown Sand up to 1 m thick locally splits the peat (Figure 23). Exposures are very restricted, for a blanket of Blown Sand covers the peat along the littoral, while inland there are substantial areas of Made Ground. Drilling shows that the Peat directly overlies the Older Marine Alluvium on which it rests at about 4.5 m above OD (Figure 23), and so is analogous to the deposits on Pilling Moss (p.58).

Blackpool – Weeton Ridge

Isolated, peat-filled depressions occur in a belt extending south-eastwards for about 8 km from Bispham. Most of these, like others elsewhere along the ridge (Longworth *in* Johnson, 1986), are probably kettle holes formed when buried masses of ground ice melted. Some of the larger ones, however, lie in an area where collapse-breccias are known in the Mercia Mudstones (p.24), and may have originated as subsidences over underlying pockets of salt. The known basins, or groups of basins, are described below, from south to north; others may remain unproven.

i   Boreholes on the line of the M55 Motorway traverse a steep-sided basin filled with up to 10 m of peat [3560 3351] (Longworth *in* Johnson, 1985). This is probably a kettle hole (Figure 20).

ii   Around Mythop [360 350], six isolated basins, all possibly salt subsidences, contain peat.

iii   An elongated area in Stanley Park contains up to 2.4 m of peat, proved by boreholes [320 361].

iv   An open space at Queenstown occupies a gentle hollow, its contours smoothed by 1.5 to 2.4 m of Made Ground (Figure 19) that conceals a probable kettle hole filled with peat to 9.45 m [3201 3672]. The boreholes show that the sides of the hollow at the margin of the peat slope at more than 1 in 3.

v   Three, oval, peat-filled hollows, about 120 m long, have been proved by drilling west of Queenstown. They have little surface expression and are too small to be shown on the 1:50 000 sheet. Their locations and peat thicknesses are: Claremont School [312 375], 1.5 m; Ribble Motors [318 369], 1.7 m together with 3.5 m of peaty clay; Gorton Street [314 366], 3.5 m.

vi   Two hollows at Brackinscal [391 379] are floored by thin peat.

vii   An elongated peat-filled hollow has been proved in many boreholes at Derby Baths [3069 3755] with up to 9.5 m of silty peat.

viii   The open space of Gynn Square overlies an elongated area where numerous boreholes have proved up to 5.90 m of peat [3087 3797] beneath Made Ground. Boreholes on its western edge show that the peaty hollow does not extend as far as the coast.

ix   A depression with up to 1.2 m of peat has been proved in borings [335 379] at School Farm.

x   A topographical depression straddles the railway line near Layton Station and is probably mostly floored by alluvium. Boreholes prove peat 3 m thick in the north-west [3220 3865], and up to 6.1 m thick beneath Made Ground in the south-east [3244 3815].

xi   A group of six hollows near Higher Moor Farm, Warbreck range from 30 to 150 m across. Four have been drilled and each contains a maximum of 4.57 m of peat [3280 3894], but thicknesses of 2.4 m are also recorded [3250 3899, 3281 3900 and 3278 3857]. In the largest of the six depressions, recent building excavations [3313 3868] uncovered an almost complete skeleton of a male elk (*Alces alces*) with two barbed points of bone (Barnes et al., 1971; Hallam et al., 1973). The artefacts are similar to those of the 'Starr Carr' Proto-Maglemosian culture of the early Mesolithic period, but pollen dating suggests a

somewhat earlier, epi-Palaeolithic age. The excavation was open in 1970 when the resurvey was in progress and the following section was then measured (for elaboration of this section and nearby shallow drill holes, see Hallam et al., 1973):

	Thickness m
Soil	0.3
Peat, dark, humic; described as carr peat by Barnes and others (1971) who noted *Betula* wood throughout, with *Phragmites* leaves and *Potamogeton* fruitstones at the base	0.30
Peat, fibrous, brown, containing sphagnum	0.10
Peat, with *Betula* wood; the elk horns reputedly protruded into the base of this peat	0.06–0.15
Clay, plastic, pale grey with rare erratic pebbles; the elk and artefacts were in this clay	0.43
Clay, dark grey, humic	0.05
Clay, plastic, grey with bivalve shells common in top 0.08 m; described as gyttya by Hallam and others (1973)	0.38
Clay	0.40

Pollen analyses (Barnes et al., 1971; Hallam et al., 1973) establish that the beds beneath the peat fall into zones I–III of the pollen zone sequence and that the elk skeleton is in zone II. Radiocarbon datings confirm this, giving ages of 11 665 ± 140 BP and 12 200 ± 160 BP (or about 10 000 years BC).

xii   Three depressions, possibly related to salt subsidence, lie adjacent to the north-east corner of the North Shore golf course. All have been proved by boring and show up to 5.33 m of peat [3184 3921].

xiii   Some gentle depressions in the built-up area of Bispham are floored by small thicknesses of alluvial silt and clay including some beds of peat. One borehole [3019 4027] near the head of one of the depressions proved 0.76 m of peat beneath 1.5 m of Made Ground and sandy clay; it was underlain by a further 1.99 m of sandy clay before boulder clay was reached. Another hole [3127 4015] passed through 1.22 m of peat beneath Made Ground; it was underlain by 0.46 m of clay separating it from a further 0.76 m of peat, lying 1.1 m above boulder clay.

xiv   A minor valley near Churchtown is floored by a similar sequence. One borehole [3152 4052] proved: fill to 1.83 m; peat to 2.13 m; gravel to 2.44 m; on clay.

xv   A much larger flat, more than 400 m long and 100 m wide, has been left as open ground [316 411] amongst the built-up area of Norbreck. Boreholes show up to 1.22 m of peat underlain by thin, silty sand and clay. It is probable that salt subsidence was responsible for the depression, in which the peat has subsequently accumulated.

xvi   A small patch of peat [3137 4242] has been proved by drilling near Anchorsholme Lane West. Only about 0.30 to 0.60 m have been recorded, separated from boulder clay by 1.5 to 3.5 m of 'marl'. The relationship of the peat to the rest of the drift is far from clear; there is Made Ground recorded here and more may have escaped recognition.

Marton Mere

Two sinuous valley systems lead from the depression containing Marton Mere; one runs northwards from Layton to Skippool, and the other trends eastwards to the Skippool Channel. Both contain scattered areas of peat, and the following thicknesses have been proved by drilling: 3.66 m [3249 3731]; 3.05 m [3279 3687]; 2.44 m [3263 3709 and 3370 3653]. Near Marton Mere, the depression is mostly floored by alluvium but some peat is preserved on its southwestern edge. Drilling in a tributary valley [3275 3522] near Stanley Park proved up to 10.67 m of peat.

Skippool Channel

The peat flooring the Skippool Channel is the most extensive spread in the district. In the south, boreholes along the M55 have proved thicknesses of up to 7 m. (Figure 20). Farther north, 1.25 m has been proved in one borehole [3679 3894], and 0.9 m can be seen at the surface overlying stoneless, grey clay. In tributary valleys, up to 1.2 m of peat are visible in several places and De Rance (1877) records thicknesses in excess of 3 m.

Coastal plain

Coastal exposures of peat within the Older Marine and Estuarine Alluvium have already been noted (pp.53,54), and De Rance recorded 'hazel nuts' in the one at Preesall Sands. Inland, there is an interesting proving beneath the tongue of Alluvium [328 414] west of Norcross. Several boreholes [e.g. 3279 4149] have reached about 1 m of thin peat or peaty clay, lying as deep as 15 m and apparently very close above boulder clay; this suggests that it may equate with the basal peat of South Shore, but its depth of almost 12 m below OD makes it possible that it formed within a depression caused by salt subsidence. Farther north, although organic bands have been noted in the Marine Alluvium, there are none of the discrete beds of peat common in other coastal locations.

East of the Wyre, there are very few provings of the Marine Alluvium to establish whether laterally persistent peat beds are present. Several boreholes at Knott End-on-Sea record thin organic bands within the silty clay, but even these have not been noted farther east.

Preesall and Hambleton

Thin organic bands occur within the Older Estuarine Alluvium inland from Barnaby's Sands (see p.55) and near Hambleton, the thickest proving being 2 m of peaty clay, with its base at 3.1 m above OD [3695 4324]. Other occurrences lie on boulder clay, their relationship to the Alluvium being less certain. One [3645 4730], occupying an area of 300 × 80 m just west of Preesall, is elongated along the strike of the Preesall Salt and almost certainly fills a salt subsidence of some antiquity; ditches expose at least 60 cm of peat beneath a skin of clay. There are two other small patches [3855 4539 and 3904 4519] near Hankinson's Farm, Stalmine, separating outcrops of Boulder Clay and Marine Alluvium.

Three small outcrops [3834 4403 to 3879 4407] on White Moss, show peat overlying Glacial Lake Deposits; these are the only remaining patches of the peat that once floored all the glacial lake sites and many of the associated drainage channels, linking to the expanse of Stalmine Moss to the

north and Pilling Moss to the east (Figure 26). Presumably, peat began to accumulate in the lakes, spread along the drainage channels and gradually encroached upon the lower slopes of till.

When the primary survey was carried out in the last century, Stalmine Moss and Pilling Moss had a thick and extensive cover of peat which rested on the surface of the Older Marine Alluvium (Figure 26). The area covered by peat has now shrunk, largely as a result of drainage, cutting and cultivation, to a small tract east of Hankinson's Farm [3913 4513]. The magnitude of the change in surface levels where the peat has gone can be seen by comparing recently determined bench-marks with the topographical contours, which have not been modified since they were first surveyed in the middle of the last century. A good example is near Wellhouse Farm [3873 4576], where the 25 ft contour, originally surveyed across the surface of the peat, runs practically through a bench-mark at 18 ft above OD just to the east of the farm where the peat has been cleared. More obviously, roads and lanes that once ran across peat spreads now stand high above the adjoining fields; the peat has gone everywhere except from beneath the roads. A good example is the farm road [3932 4591] east of Wellhouse Farm, where the track runs on a causeway of peat, too narrow to show on the 1:50 000 sheet, rising about 1.5 m above the surface of the Alluvium. Even within much of the remaining area of peat, the surface has fallen for, immediately to the east of the district, both the road leading eastwards from Hankinson's Farm, and Boundary Lane, now a field-track leading southwestwards from The Woodlands, stand well above the fields that they cross. Only in uncleared patches of woodland, as at Tarn Wood [3979 4583], does the present surface approximate to that of mid-Victorian times.

The stratigraphical position of the peat is unambiguous. It clearly rests on the Older Marine Alluvium in the north and covers the ridges of the Older Storm Beaches to rest on till southwards from Hankinson's Farm. If it spread northwards from the floors of the old glacial lakes, then its base is presumably diachronous, becoming steadily younger northwards. The preserved remnants are, however, among the youngest peats in the region. By analogy with the district to the east (Oldfield and Statham, 1965), the base of the peat, where it overlies the almost planar surface of the Marine Alluvium, is likely to lie within pollen zone VII; radiocarbon analyses show it to date from about 5000 BP. Investigations of the later vegetational history of the mosslands, complicated by the steadily increasing effects of human activity, are summarised by Barnes (1975).

Blown Sand

Deposits of Blown Sand extend along practically the entire coast, except between Norbreck and the North Pier and from Rossall School to Fleetwood golf links.

From the Central Pier to South Shore there is an extensive sand sheet, mostly covered by houses. Exposures are limited to the slopes of the former railway marshalling yards [3110 3440], where 0.9 m of fine- to medium-grained sand is visible, but the best records are from the many shallow boreholes. They show that the sand is generally more than 2.5 m thick and attains 4 m locally (Figure 23). Towards

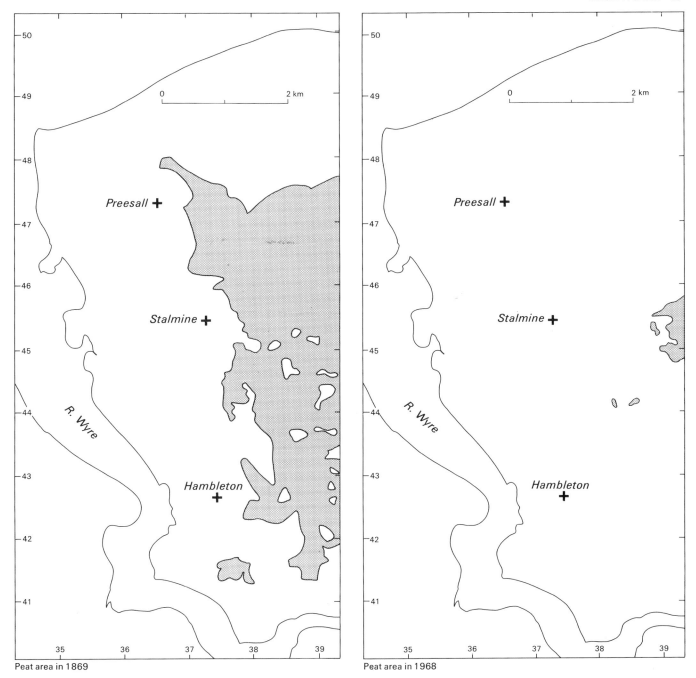

Figure 26 Former and present extent of peat east of the River Wyre

Central Pier the sand is less continuous and the underlying peat is split locally by a sand lens up to 0.8 m thick which may also be wind-blown (Figure 23).

The top of the cliff at Blackpool is without Blown Sand, since it is probably too high for wind to carry up the sand. The deposit begins again near Norbreck, though as far north as a point [3116 4160] west of Little Bispham, there is some doubt as to whether it is a natural deposit. Up to 3 m of pale sand can be seen resting upon boulder clay, but it lies at the top of a substantial cliff and it is possible that it was placed there when early coast protection works were being carried out.

There can be no such reservation about the narrow strip of Blown Sand that continues northwards to Rossall School [313 449] though the original surface is much modified. The deposit rests on a Storm Beach along much of its length, though this is not everywhere apparent. At the inland margin of the Blown Sand [3136 4390], trenches have been dug through into underlying Marine Alluvium. Northwards, there are short lengths of Storm Beach preserved along the Promenade but the absence of associated Blown Sand makes it likely that the beach is recent, and that any older beaches or dunes were destroyed by a substantial breaching of the barrier they once formed.

The Blown Sand reappears [311 470] a little to the south of Rossall Scar, and continues eastwards to the mouth of the Wyre to form the largest outcrop in the district. Around Rossall Point it again rests upon a Storm Beach, and the number of stones dug from graves in Fleetwood cemetery [319 478] makes it possible that the Blown Sand rests on Glacial Sand or on a shingle bar comparable with the Preesall Shingle, though the entire area has been too much disturbed by human activity to confirm this. Probably the least modified area is The Mount [3338 4833], the nodal point of the 'new town' of the early 19th century, where dunes still stand to about 16 m above OD.

The line of dunes ends abruptly at the Wyre and, except for a small tract [351 453] on a north-west-facing shore at Heads, there is no Blown Sand along the Wyre estuary. The Blown Sand begins again at Knott End-on-Sea, though its outcrop has been heavily modified by coastal protection works and, from Cocker's Dyke Houses [377 494] eastwards, by reclamation schemes. It is impossible in places to differentiate between genuine Blown Sand and artificial sand emplacements; there are no obvious dunes and a locally extensive, though thin, wash of sand further serves to obscure the inner edge of the Blown Sand, as does the presence of an underlying bar of sand in the Older Marine Alluvium (p.55).

The Blown Sand terminates abruptly [3944 4988] just to the east of Fluke Hall where, if it were not confined by embankments, it would be encroaching on the muddy salt-marsh to the east.

Alluvium and Beach deposits

Several small spreads of contemporary fluvial alluvium flank small streams that debouch onto the contemporary or Older Marine and Estuarine Alluvium, [e.g. 352 466]. Most are too restricted to be indicated on the 1:50 000 sheet, but are shown on the 1:10 000 maps. Their distinction is largely arbitrary, particularly where flood-gates limit the inward flow of the tides, and it is likely that the freshwater clays and silts are underlain by several metres of Estuarine Alluvium that have filled ancient creeks [e.g. 356 404]. Other small strips floor some of the old kettle holes and glacial channels on the Blackpool-Weeton ridge [e.g. 340 377]. Between Layton and Poulton, Horse Bridge Watercourse runs along a sinuous depression floored by Alluvium. Several boreholes prove variable deposits of clay, sand and gravel, ranging up to 6.1 m in depth. In places, patches of peat blanket these. Peaty clay has also been dug in the floors of small valleys tributary to the Skippool Channel.

Contemporary Marine and Estuarine Alluvium is practically restricted to the salt-marsh [395 505] east of Fluke Hall and to the banks of the Wyre. To the west of the river, an extensive salt-marsh [335 460] has been almost entirely reclaimed, and much of Wyre Water, shown on the 1st edition of the Ordnance Survey one-inch sheet, has been filled and built over. Farther south, a more or less continuous strip of chocolate-brown and purple-grey mud makes up the saltings from Stanah to about 1 km east of Shard Bridge. On the east side of the river, the biggest expanse of salt-marsh lies near Barnaby's Sands, where it is accreting to the east of the low drumlin forming Arm Hill [3470 4630] and continues southwards towards Heads. There are smaller spreads of saltings at Burrow's Marsh [354 450], for about 880 m upstream from Wardley's Pool [3665 4280], and on either side of Shard Bridge. This last strip expands in the extreme east and appears to pass into the river alluvium of the Wyre. Minor steps in its surface seem to have resulted from erosion during tidal surges, and the deposit changes upstream to a brown, silty clay that is laminated by thin partings of silt.

There are few signs of contemporary Storm Beach, largely because the coastal defences are so extensive, though there are narrow strips for about 400 m on either side of Rossall Point. The contemporary Beach Deposits are dominantly clean sand, with sporadic strips of beach shingle near high-water mark northwards from Cleveleys, and others parallel to the coast but about 100 m seaward of it. At Rossall Scar, where the coast changes direction, there is a wider expanse of beach shingle, which feeds cobble patches along the north-facing coast where the intertidal beach is up to 3 km wide. East of the Wyre, apart from an area of cobbles at Great Knott [340 490], the Beach Deposits are wholly of sand and extend for about 4.5 km northwards to the Lune channel. Only in the extreme east near Fluke Hall do the sands give way to mud-flats.

CHAPTER 6

Economic geology

ROCK-SALT AND BRINE

Salt has been won in the western Fylde since medieval times, when there were brine-pans on the coastal marshes near Pilling. The modern industry, however, dates back to boreholes fortuitously drilled around Preesall in a misconceived search for hematite. One of the holes proved a bed of rock-salt, and by 1885 a shaft [3630 4672] sunk near its site had penetrated over 100 m of the evaporite. In 1889 the Preesall Salt Co. Ltd began to pump brine from the shaft, and erected a saltworks [338 449] alongside the railway about 3 km south of Fleetwood, which was fed with Preesall brine through a pipeline running beneath the Wyre. In the following year, the company became part of the larger United Alkali Co. Ltd; this firm promptly built an ammonia-soda plant [342 442] a little to the south of the salt-works and sank two new shafts [3602 4661 and 3605 4663] from which rock-salt was soon mined. With the demise of the salt-works, the main user of the brine became the ammonia-soda plant, while rock-salt was shipped out especially to the company's other works on Merseyside and in Flint, Scotland and Bristol. In 1926 the United Alkali Co. Ltd was one of the founding component companies of Imperial Chemical Industries Ltd, who modernised and expanded the working brine-field — the mine was abandoned early in the 1930's — to meet a changing chemical market. The brine is now the feed-stock for a modern chemical plant at Hillhouse Works, which replaced the nearby ammonia-soda plant when this became obsolescent. The new plant uses the Castner-Kellner process as the starting point for the manufacture of PVC, polyurethane and bleach.

Government production figures chart the early development of the Preesall field, for, before 1919, they appeared on a county basis and Preesall was the only Lancashire producer; they first record salt-in-brine tonnages in 1890, and rock-salt output from the mine is first noted in 1892. After 1919, the published salt-in-brine figures are included in a single figure that includes the output from Staffordshire, Yorkshire and the Isle of Man, so they are valueless as a guide to Preesall brine output. The Preesall mine was, however, the only producer of rock-salt in the region and, consequently, the published figures relate solely to it. The workings were at two levels, each taking about a 6 m section. One set of workings was within A-Bed, and the other was high in C-Bed (see Figure 8 and Plate 16) about 13 m below the top of the sequence. At this upper level lenticles of anhydrite, the thickest about 25 cm thick, are said to have occurred near the top of the mined section. The area of mining was not large, amounting to about 350 × 500 m. The latest recorded rock-salt production was in 1930 and soon after this the mine was abandoned, though it was used in wartime as an underground brine reservoir. The relevant production statistics, converted to tonnes, are shown in Figure 27. The current published Government figures are of little use in determining production from individual fields, but Imperial Chemical Industries Ltd have kindly helped by providing additional statistics.

Even though production from Preesall has always been comparatively small, the field is of particular interest in the part it played in the development of the technique of controlled brine-pumping that now dominates the industry in Britain. From its earliest days, the Preesall field suffered from a wholly inadequate supply of natural brine, a result of the thick and complete cover of impermeable boulder clay that covers the wet-rockhead area. Because of this, the Fleetwood Salt Co. Ltd drilled boreholes around their pumping shaft and fed water down them to augment the natural supply. This procedure was improved upon by the United Alkali Co. Ltd, who used two strings of piping in each hole, one within the other. Fresh water was passed down between the two pipes, and the resultant brine was pumped up the inner pipe, which penetrated much the deeper of the two and tapped the brine that collected in the bottom of the growing solution cavity. At first, these cavities grew until they eventually collapsed, sometimes with dramatic subsidence at the surface. Over the years, however, the company and its ICI successor evolved techniques of maintaining the pressure-tightness of the cavities, both during use and after their demise. As a result, surface subsidence has become a thing of the past.

The future of the saltfield depends wholly on the future of the Hillhouse Works; while that thrives, so will Preesall. A long-term limitation is, however, imposed by the restricted geographical limits of the field. Both to the east and south, the salt comes to crop, while new housing developments to the north have sterilised a further area. The western limits of the field remain largely unexplored on land, and wholly so beneath the Wyre; assessment of the long-term prospects in this direction depends on the detailed geological structure which is still uncertain. Two further possibilities remain for continuing production locally when the present field is exhaused. First, the salt probably extends northwards offshore beneath Morecambe Bay. Second, it is possible that Permian salt underlies the Triassic salt in the core of the Preesall Syncline; salt of this age occurs in the offshore gasfield. However, the depth at which it might lie would be a severe constraint on its commercial exploitation.

UNDERGROUND WATER

In the north-east of the district, several wells were sunk into the Sherwood Sandstone to supply fresh water to the Preesall saltfield. The yields were substantial and some of the boreholes are still in use up to the present day. Two boreholes, now disused, were also sunk into these beds by Blackpool Corporation Electricity Department.

The Mercia Mudstones were first tested for water supply

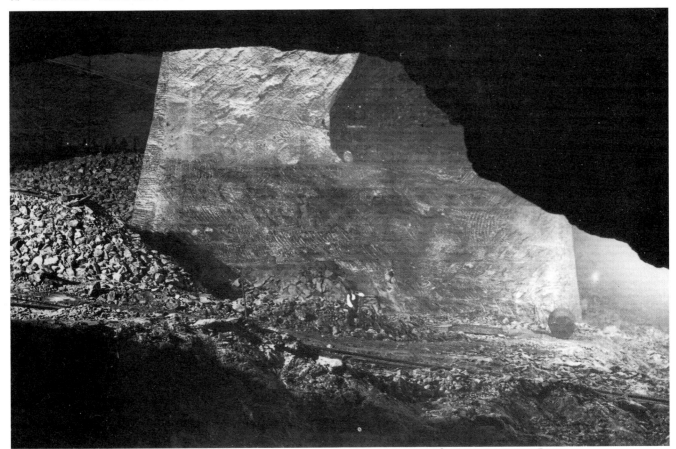

Plate 16   The former rock-salt mine at Preesall. View of the Top Mine, supported by a pillar 18 m in width (*Imperial Chemical Industries plc*)

at Fleetwood Barracks, now the site of the North Euston Hotel [3378 4844]. The borehole, drilled to a total depth of 170.40 m was found to be dry. A borehole, now disused, drilled in 1905 at the Blackpool Winter Gardens is thought to have yielded only saline water, derived from the halite at the bottom of the hole.

Boreholes into Middle Sands have supplied water to several market gardens, a brewery, gasworks and small factories, but most of the boreholes are now disused. One of the best yields, 4200 gallons per hour under continuous pumping, was obtained at Marton gasworks [3451 3369] from 44 m of Middle Sand, although the water was mineralised and undrinkable, with 450 ppm of sulphates and 1800 ppm of chlorides. The water was probably derived indirectly from Mercia Mudstones which are thought to occur at the bottom of the borehole.

A borehole at Thornton le Fylde gasworks [3408 4370], terminating at 46 m in glacial sand, was never used because the water was brackish. This hole is near both the outcrop of the Preesall Salt and the tidal waters of the River Wyre.

BRICK CLAY

Brick clay was formerly obtained from two brick-pits in the Upper Boulder Clay in Blackpool. Eaves Bros. pit on Rec-

tory Road [330 341] was working at the time of survey, but has now closed. The working face was 6 m high. Common bricks were manufactured after the removal of erratics from the till. Warbeck Hill Brickworks, formerly worked to a depth of 6 m, has been closed for many years. Much of Blackpool is brick-built, but the bulk of the brick has been brought in from East Lancashire. In particular, Accrington Red bricks have been extensively used.

SAND AND GRAVEL

At the time of the resurvey, there were two diggings for building sand at Limebrest Farm sand-pit [3470 4171], just south of Thornton. At least 3.7 m of sand were exposed, but further expansion is severely restricted by surrounding housing. Building sand was also being worked in Greaves Bros. pit at Hardhorn [351 373], where 2 m of sand were being worked under a cover of up to 5 m of Upper Boulder Clay, but the working is now abandoned. Sand was once worked on a small scale in two pits, at Singleton Windmill [383 377] and at Preese Hall [376 361]. During the construction of the M55 Motorway, advantage was taken of the easily accessible outcrop of sand [3886 3360] flanking the deep cuttings in both the active and defunct railway lines at Great Plumpton to extract sand with scattered pebbles for the motorway

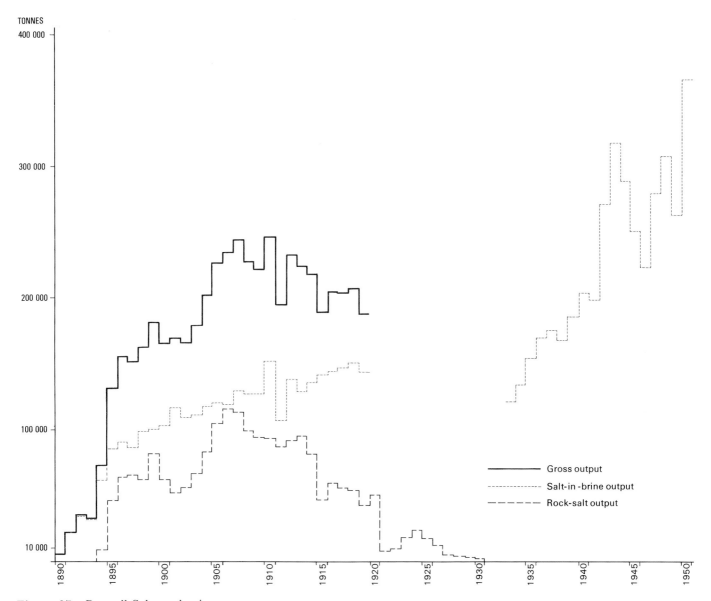

Figure 27   Preesall Salt production

works.

Gravel was once dug from the foreshore near Blackpool and Fleetwood, but the practice did not persist for long because of its effect on coastal erosion. The only significant inland workings are in the Preesall Shingle, east of Knott End-on-Sea. There used to be at least 10 gravel pits, and probably others which have been filled-in and so are now dry. Future prospects are poor, for the deposit is limited in size and housing is steadily encroaching around its outcrop. The other ridges of Older Storm Beach are likely to be too small to maintain a modern pit, though they provide a useful local supply to farms.

ENGINEERING GEOLOGY

The solid formations are deeply buried beneath Quaternary deposits and do not feature in this account. Site investigation is particularly appropriate on the extensive tracts of silty Marine Alluvium, in places overlain by peat, and also on the boulder clay terrain where peat-filled kettle holes have proved a major local hazard. Some 4000 borehole records, chiefly site-investigation and sewerage boreholes, are stored in the BGS National Geosciences Data Centre. The bulk of them, about 3000, are also curated by the Department of the Borough Engineer at Blackpool which has kept an efficient database for more than 30 years. The properties of the deposits are described below in ascending stratigraphical order.

Lower Boulder Clay

This is not present in foundations except locally along the shore, where it has been encountered in the coastal defence works at the base of the Bispham cliffs and in the sewerage outfall at Anchorsholme, Cleveleys.

When the Anchorsholme sewer outfall was being constructed, a ridge of very hard boulder clay, almost certainly the Lower Boulder Clay, obstructed digging for 130 m. Seven samples from three trial pits gave an average, undrained shear strength in the triaxial test of 247.8 kPa, the highest value being 412.6, the lowest 88.4. This compares with the average for the Upper Boulder Clay of 135.3 kPa (see below).

During excavations for the sea wall between 1917 and 1921, 'on account of the rock-like toughness of the boulder clay gelignite was used to facilitate and cheapen the cost of getting out the trench' (Banks, 1936). Recently steel piles were installed at the base of the sea wall, but in places only short 2 m piles could be driven due to the strength and stiffness of the deposit.

Middle Sands

These are rarely seen at surface, but have been tested at depths of 10 to 25 m in several site-investigation boreholes. There is a wide spread of values in the Standard Penetration Test from 10 to 133, averaging 47 (30 samples). This indicates the variation in density to be expected in this deposit.

Upper Boulder Clay

Most of Blackpool, excepting South Shore, is built on Upper Boulder Clay. This makes a good foundation and the most notable building, resting on about 16 m of Upper Boulder Clay, is the 158 m-high Blackpool Tower (Figure 19, section A–B–C). Detailed site investigations are usually specified by the authorities for high structures in the boulder clay terrain in order to detect peat-filled kettle holes. Since peat is compressible when loaded, it is a threat to foundations.

An average, undrained, triaxial shear strength test on borehole cores through the Upper Boulder Clay gives 135.3 kPa, the lowest value being 36, the highest 226 (40 samples). Standard Penetration Tests averaged 92 blows, the highest being 150 and the lowest 56, indicating the moderate stiffness of the deposit.

Marine Alluvium of Flandrian age

Between Fleetwood and Cleveleys, where these deposits occur widely, 35 Standard Penetration Test values showed an average of 8.5, with a range of 4 to 30 blows. In most cases, the value rose greatly in the basal 1 to 2 m of the deposit and if one ignores the one reading taken in each borehole closest to the base of the deposit (average of 29 tests), the average value is only 6.6. The gravels and sands at the base of the sequence (Figure 24) commonly contain water under artesian pressure.

On Blackpool South Shore, the Marine Alluvium is commonly too deeply buried under peat and blown sand to form the immediate foundation. However, the silty clays and silty sands at 2 to 6 m deep are penetrated by the foundations of large structures and may necessitate piled or vibroflotation foundations. The deep sewers in Blackpool have been difficult to drive through the Marine Alluvium, mainly because the silty clays which are dominant in the upper layers of the Flandrian marine alluvium are not easy to dewater. Standard Penetration Tests on 20 samples averaged 8.2, with a range of 3 to 27 blows.

Within the Marine Alluvium on South Shore is an old beach deposit of sand and gravel. Standard Penetration Tests on 23 samples from this averaged 39.6, with a range of 12 to 57 blows, showing the deposit to be significantly more dense than the silty beds.

Lacustrine Deposits

These are the silts, clays and thick peats found locally in kettle holes and the extensive deposits around Marton Mere. Boreholes in the flats around the lake are in very soft clays to a depth of 9.8 m, with a range from 1 to 2 blows (15 samples), in Standard Penetration Tests.

Peat

One of the greatest problems for construction in Blackpool is presented by isolated kettle holes, in which compressible peat is up to 8 m thick and locally saturated with water (Figure 19). Several such areas have been identified by earlier builders and left undeveloped, as at Queenstown Recreation Ground and Gynn Square. Other hollows have been partially developed by using deep friction piling penetrating into the underlying boulder clay (19 m maximum pile length). Where these precautions have not been taken, there has been subsidence, even of light structures like commercial greenhouses, and in the past some dwellings have had to be dismantled.

On Blackpool South Shore, peat is present almost everywhere beneath Blown Sand and varies in the degree to which it has been previously compressed. The extensive terraces of Edwardian houses have basements dug into the Blown Sand in which the cellar floors are close to the top of the underlying peat. Commonly, the walls of each house have sunk farther into the compressible peat than the cellar floor which has therefore developed a marked central hump. In cases where this hump has been removed, it has re-formed after a period of years. Nevertheless, there has been little serious damage from this gradual subsidence. Recent building on the South Shore has been chiefly of larger structures, and preliminary site investigation has been followed by installation of piled or vibroflotation foundations.

HYDROCARBON POTENTIAL

About 35 km west of Blackpool lies the commercially successful Morecambe Bay gasfield. It is situated in the southern part of the Keys Basin, a sub-basin of the main East Irish Sea Basin, where there are substantial gas-traps in the Sherwood Sandstone beneath a capping of Mercia Mudstone. It is possible that there are further traps within the Blackpool district, though these must be much shallower.

For example, one possiblity is the Weeton Anticline where there has been extensive hydrocarbon reduction of the red ferric cement of the Sherwood Sandstone, but the present cover is quite thin (see Figure 15). It may be thicker in the upfaulted area lying between the Preesall Syncline and the Formby Point Fault, beneath the sea about 4 km west of Blackpool. This area seems to be a northerly continuation of the structural block containing the small Formby oilfield, though the total depth of burial is almost certainly much less than in the Keys Basin, and platform areas of this sort are not normally promising sites for the accumulation of hydrocarbons.

If any hydrocarbons are present around Blackpool, they are likely to have migrated upwards in Mesozoic times from Carboniferous source rocks at depth. Both Dinantian and Namurian successions contain likely source rocks, such as the Worston and Bowland Shales. The best local source, however, is likely to be the Coal Measures within the Westphalian succession, for the collieries around Whitehaven were notoriously gassy and Point of Air Colliery, on the coast of Clywd, has yielded sufficient methane to be worth piping to the surface. Whether Westphalian rocks underlie the district is not, however, known.

Potential reservoir rocks other than the Sherwood Sandstone include the grits in the Namurian sequence and the Collyhurst Sandstone within the Permian. Again the presence of these beds is, as yet, speculative.

REFERENCES

Most of the references listed below are held in the Library of the British Geological Survey at Keyworth, Nottingham. Copies of the references can be purchased subject to the current copyright legislation.

AUDLEY-CHARLES, M G. 1970. Stratigraphical correlation of the Triassic rocks of British Isles. *Quarterly Journal of the Geological Society of London*, Vol. 126, 19–47.

BANKS, H. 1936. Blackpool coast defence works. 94–103 in *A scientific survey of Blackpool and district*. GRIME A (editor). (London: British Association.)

BARKER, R D. 1974. A gravity survey of northwest Lancashire. *Geological Journal*, Vol. 9, 29–38.

BARNES, B. 1975. Palaeoecological studies of the late Quaternary period in northwest Lancashire. Unpublished PhD thesis, University of Lancaster.

— EDWARDS, B J N, HALLAM, J S, and STUART, A J. 1971. Skeleton of a Late Glacial elk associated with barbed points from Poulton-le-Fylde, Lancashire. *Nature, London*, Vol. 232, 488–489.

BINNEY, E W. 1852. Notes on the drift deposits found near Blackpool. *Memoirs of the Literary and Philosophical Society of Manchester*, Vol. 10, Ser. 2, 121–135.

BOULTON, G S. 1972. The role of thermal regime in glacial sedimentation. 1–19 in Polar geomorphology. PRICE, R J, and SUGDEN, D E (editors). *Special Publication Institute of British Geographers*, No. 4.

BRADBURY, E E. 1971. Palynological and stratigraphic investigations in the Skitham area of Rawcliffe Moss. Unpublished BEd dissertation, Poulton-le-Fylde College of Education.

BRUGMAN, W A. 1986. Late Scythian and Middle Triassic palynostratigraphy in the Alpine realm. *Albertiana*, Vol. 5, 19–20.

BURGESS, W G, WHEILDON, J, GEBSKI, J S, SARTORI, A, WILSON, A A, and FROST, D V. 1984. *Heat flow measurements in north-west England. Investigation of the geothermal potential of the UK.* (Keyworth: British Geological Survey.)

CARRUTHERS, R G. 1939. On northern glacial drifts; some peculiarities and their significance. *Quarterly Journal of the Geological Society of London*, Vol. 95, 299–333.

COLTER, V S, and BARR, K W. 1975. Recent developments in the geology of the Irish Sea and Cheshire Basins. 61–73 in *Petroleum and the continental shelf of northwest Europe. I. Geology.* WOODLAND, A W (editor). (London: Applied Science Publishers.)

DE RANCE, C E. 1875. The geology of the country around Blackpool, Poulton and Fleetwood. *Memoir of the Geological Survey of Great Britain.* 14 pp.

— 1877. The superficial geology of the country adjoining the coasts of southwest Lancashire. *Memoir of the Geological Survey of Great Britain.* 139 pp.

EARP, J R, and TAYLOR, B J. 1986. Geology of the country around Chester and Winsford. *Memoir of the Geological Survey of Great Britain.* 119 pp.

EEM, J G L A, VAN DER. 1983. Aspects of middle and late Triassic palynology. 6. Palynological investigations in the Ladinian and lower Karnian of the Western Dolomites, Italy. *Review of Paleobotany and Palynology*, Vol. 39, 189–300.

ELLIOTT, R E. 1961. The stratigraphy of the Keuper Series in southern Nottinghamshire. *Proceedings of the Yorkshire Geological Society*, Vol. 33, 197–234.

EVANS, W B. 1970. The Triassic salt deposits of north-western England. *Quarterly Journal of the Geological Society of London*, Vol. 126, 103–123.

— and WILSON, A A. 1975. Outline of geology on Sheet 66 (Blackpool) of 1:50 000 Series: Geological Survey of Great Britain.

— — TAYLOR, B J, and PRICE, D. 1968. Geology of the country around Macclesfield, Congleton, Crewe and Middlewich (2nd edition). *Memoir of the Geological Survey of Great Britain.* 328 pp.

EYRE, K. 1961. *Seven Golden Miles, the fantastic story of Blackpool.* 170 pp. (Lytham St. Annes: Weaver and Youles.)

FISHER, M J. 1972. The Triassic palynofloral succession in England. *Geoscience and Man*, Vol. 4, 101–109.

FORSTER, S C, and WARRINGTON, G. 1985. Geochronology of the Carboniferous, Permian and Triassic. 99–113 in The chronology of the geological record. SNELLING, N J (editor). *Memoir of the Geological Society of London*, No. 10.

FROST, D V, and SMART, J G O. 1979. Geology of the country north of Derby. *Memoir of the Geological Survey of Great Britain.* 199pp.

GRESSWELL, R K. 1953. *Sandy shores in south Lancashire. The geomorphology of south-west Lancashire.* (Liverpool: Liverpool University Press.)

— 1957. Hillhouse coastal deposits of south Lancashire. *Liverpool and Manchester Geological Journal*, Vol. 2 60–78.

— 1967. The geomorphology of the Fylde. 25–42 in *Liverpool essays in geography: a Jubilee collection.* STEEL, R W, and LAWTON, R, (editor). (London: Longman.)

HALLAM, J S, EDWARDS, B J N, BARNES, B, and STUART, A J. 1973. The remains of a Late Glacial elk associated with barbed points from High Furlong near Blackpool, Lancashire. *Proceedings of the Prehistorical Society*, Vol. 39, 100–128.

IRELAND, R J, POLLARD, J E, STEEL, R J, and THOMPSON, D B. 1978. Intertidal sediments and trace fossils from the Waterstones (Scythian–Anisian?) at Daresbury, Cheshire. *Proceedings of the Yorkshire Geological Society*, Vol. 41, 399–436.

JACKSON, D I, MULHOLLAND, P, JONES, S M, and WARRINGTON, G. 1987. The geological framework of the East Irish Sea Basin. 191–203 in *Petroleum geology of North-west Europe.* BROOKS, J, and GLENNIE, K (editors). (London: Graham and Trotman.)

JONES, R C B, TONKS, L H, and WRIGHT, W B. 1938. The geology of the country around Wigan. *Memoir of the Geological Survey of Great Britain.* 244 pp.

LONGWORTH, D. 1985. The Quaternary history of the Lancashire Plain. 178–200 in *The geomorphology of North-West England*, JOHNSON, R H (editor). (Manchester: Manchester University Press.)

MACKINTOSH, D. 1869. On the correlation, nature and origin of the drifts of north-west Lancashire and a part of Cumberland, with remarks on denudation. *Quarterly Journal of the Geological Society of London*, Vol. 25, 407–431.

MOSTLER, H, and SCHEURING, B W. 1974. Mikrofloren aus dem Langobard und Cordevol der Nördlichen Kalkalpen und das Problem des Beginns der Keupersedimentation im Germanischen Raum. *Geologisches-Paläontologisches Mitteilungen, Innsbruck*, Vol. 4, 1–35.

OLDFIELD, F. 1956. The mosses and marshes of north Lancashire. Unpublished BA thesis, Liverpool University.

— and STRATHAM, D C. 1965. Stratigraphy and pollen analysis on Cockerham and Pilling Mosses, North Lancashire. *Memoirs and Proceedings of the Manchester Literary and Philosophical Society*, Vol. 107, 1–16.

ROSE, W C C, and DUNHAM, K C. 1977. Geology and hematite deposits of South Cumbria. *Memoir of the Geological Survey of Great Britain*. 179 pp.

— 1974b. Palynology of the Triassic. 19–21, 105 *in* Geology of the country around Stratford-upon-Avon and Evesham. WILLIAMS, B J, and WHITTAKER, A. *Memoir of the Geological Survey of Great Britain*. 127pp.

— 1974c. Les évaporites du Trias britannique. *Bulletin de la Société Géologique de la France*, (7), Vol. XVI, 708–723.

— 1978a. Palynology of the Triassic sequence in the Port More Borehole. 76–77, 101 *in* Geology of the Causeway Coast. WILSON, H E, and MANNING, P I. *Memoir of the Geological Survey of Northern Ireland*. 172pp.

— 1978b. Palynology of the Keuper, Westbury and Cotham beds and the White Lias of the Withycombe Farm Borehole. *Bulletin of the Geological Survey of Great Britain*, No. 68, 22–28.

— 1982. Palynological studies of the Triassic succession in the Ashbourne district. 15–18 *in* A standard nomenclature for the Triassic formations of the Ashbourne district. CHARSLEY, T J. *Report of the Institute of Geological Sciences*, No. 81/14.

— 1986. Palynology. 34 *in* Geology of the country around Chester and Winsford. EARP, J R, and TAYLOR, B J. *Memoir of the British Geological Survey*. 119pp.

— 1988. Mercia Mudstone palynology. 67 *in* Geology of the country around Coalville. WORSSAM, B C, and OLD, R A. *Memoir of the British Geological Survey*. 161pp.

— AUDLEY-CHARLES, M G, ELLIOTT, R E, EVANS, W B, IVIMEY-COOK, H C, KENT, P E, ROBINSON, P L, SHOTTON, F W, and TAYLOR, F M. 1980. A correlation of Triassic rocks in the British Isles. *Special Report of the Geological Society of London*, No. 13. 78 pp. (Oxford: Blackwell Scientific Publications.)

WRAY, D A, and COPE, F W. 1948. Geology of Southport and Formby. *Memoir of the Geological Survey of the United Kingdom*. 54 pp.

SCHEURING, B W. 1970. Palynologische und palyno-stratigraphische Untersuchungen des Keupers im Bölchentunnel (Solothurner Jura). *Schweizerische Paläontologische Abhandlungen*, Vol. 88, 1–119.

SHEARMAN, D J. 1970. Recent halite rock, Baja California, Mexico. *Transactions of the Institute of Mining and Metallurgy*, Vol. 79, B155–B162.

SHERLOCK, R L. 1921. Special reports on the mineral resources of Great Britain. Volume XVIII — Rock-salt and brine. *Memoir of the Geological Survey of Great Britain*. 122 pp.

SHOTTON, F W, BLUNDELL, D J, and WILLIAMS, R E G. 1970. Birmingham University radiocarbon dates IV. *Radiocarbon*, Vol. 12, No. 2, 385–399.

SMITH, E G, and WARRINGTON, G. 1971. The age and relationships of the Triassic rocks assigned to the lower part of the Keuper in north Nottinghamshire, north-west Lincolnshire and south Yorkshire. *Proceedings of the Yorkshire Geological Society*, Vol. 38, 201–227.

TAYLOR, B J. 1958. Cemented shear-planes in the Middle Sands of Lancashire and Cheshire. *Proceedings of the Yorkshire Geological Society*, Vol. 31, 359–365.

THOMPSON, D B. 1970a. Sedimentation of the Triassic (Scythian) Red Pebbly Sandstones in the Cheshire Basin and its margins. *Geological Journal*, Vol. 7, 183–216.

— 1970b. The stratigraphy of the so-called Keuper Sandstone Formation (Scythian – ?Anisian) in the Permo-Triassic Cheshire Basin. *Quarterly Journal of the Geological Society of London*, Vol. 126, 151–181.

THOMPSON, F J. 1908. The rock-salt deposits at Preesall, Fleetwood, and the mining operations therein. *Transactions of the Manchester Geological and Mining Society*, Vol. 30, 105–116.

THORNBER, W A B. 1837. *An historical and descriptive account of Blackpool and it's neighbourhood*. 352 pp. (Poulton, Thornber W A B.) Reprinted 1985, 65 pp. (Blackpool: Blackpool and Fylde Historical Society.)

TOOLEY, M J. 1969. Sea level changes and the development of coastal plant communities during the Flandrian in Lancashire and adjacent areas. Unpublished PhD thesis. University of Lancaster.

— 1971. Changes in sea-level and the implication for coastal development. *In* Association of river authorities yearbook and directory, 1971.

— 1976. Flandrian sea-level changes in west Lancashire and their implications for the Hillhouse Coastline. *Geological Journal*, Vol. 11, part 2, 137–152.

TUCKER, R M. 1981. Giant polygons in the Triassic salt of Cheshire, England: a thermal contraction model for their origin. *Journal of Sedimentary Petrology*, Vol. 51, 779–786.

VISSCHER, H, and BRUGMAN, W A. 1981. Ranges of selected palynomorphs in the Alpine Triassic of Europe. *Review of Palaeobotany and Palynology*, Vol. 34, 115–128.

WARRINGTON, G. 1970a. The stratigraphy and palaeontology of the 'Keuper' Series of the central Midlands of England. *Quarterly Journal of the Geological Society of London*, Vol. 126, 183–223.

— 1970b. Palynology of the Trias of the Langford Lodge Borehole. 44–52 *in* Geology of Belfast and the Lagan Valley. MANNING, P I, ROBBIE, J A, and WILSON, H E. *Memoir of the Geological Survey of Northern Ireland*. 172pp.

— 1974a. Studies in the palynological biostratigraphy of the British Trias. I. Reference sections in west Lancashire and north Somerset. *Review of Palaeobotany and Palynology*, Vol. 17, 133–147.

— 1974b.   Palynology of the Triassic.   19–21, 105 *in* Geology of the country around Stratford-upon-Avon and Evesham. WILLIAMS, B J, and WHITTAKER, A.   *Memoir of the Geological Survey of Great Britain.* 127pp.

— 1974c.   Les évaporites du Trias britannique.   *Bulletin de la Société Géologique de la France*, (7), Vol. XVI, 708–723.

— 1978a.   Palynology of the Triassic sequence in the Port More Borehole.   76–77, 101 *in* Geology of the Causeway Coast. WILSON, H E, and MANNING, P I.   *Memoir of the Geological Survey of Northern Ireland.* 172pp.

— 1978b.   Palynology of the Keuper, Westbury and Cotham beds and the White Lias of the Withycombe Farm Borehole. *Bulletin of the Geological Survey of Great Britain*, No. 68, 22–28.

— 1982.   Palynological studies of the Triassic succession in the Ashbourne district.   15–18 *in* A standard nomenclature for the Triassic formations of the Ashbourne district. CHARSLEY, T J. *Report of the Institute of Geological Sciences*, No. 81/14.

— 1986.   Palynology.   34 *in* Geology of the country around Chester and Winsford. EARP, J R, and TAYLOR, B J.   *Memoir of the British Geological Survey.* 119pp.

— 1988.   Mercia Mudstone palynology.   67 *in Geology of the country around Coalville*.   WORSSAM, B C, and OLD, R A.   *Memoir of the British Geological Survey.* 161pp.

— AUDLEY-CHARLES, M G, ELLIOTT, R E, EVANS, W B, IVIMEY-COOK, H C, KENT, P E, ROBINSON, P L, SHOTTON, F W, and TAYLOR, F M. 1980.   A correlation of Triassic rocks in the British Isles.   *Special Report of the Geological Society of London*, No. 13. 78 pp. (Oxford: Blackwell Scientific Publications.)

WRAY, D A, and COPE, F W. 1948.   Geology of Southport and Formby.   *Memoir of the Geological Survey of the United Kingdom.* 54 pp.

APPENDIX 1

Selected boreholes

A selection of the stratigraphically most important boreholes is listed alphabetically below. Large portions of many of these summary logs are given in extra detail in the stratigraphical figures to which they are cross-referenced. The registered number of the record in the National Geosciences Data Centre at Keyworth is given after the National Grid reference.

All formations underlying the glacial deposits are of Triassic age and, apart from specified parts of boreholes in the Sherwood Sandstone Group, all remaining formations are in the overlying Mercia Mudstone Group. In order to save space, the label 'Mercia Mudstone Group' has therefore been omitted.

B1 Borehole [3390 4479] SD34SW/2

5.5 m above OD. See Figures 5 and 6. Drilled in 1958 for Imperial Chemical Industries. Cored from 18.29 to 473.58 m. Logged by B J Taylor. Dip averages 14°.

	Thickness m	Depth m
MADE GROUND		
Ashes	2.44	2.44
MARINE ALLUVIUM		
Silt, sandy	3.96	6.40
GLACIAL DEPOSITS		
Clay and gravel	2.13	8.53
Sand and gravel	2.14	10.67
Clay and gravel	3.96	14.63
Clay, sandy	3.66	18.29
THORNTON MUDSTONES		
Mudstone, red	12.80	31.09
Mudstone, grey alternating with red, laminated at least in part	97.84	128.93
SINGLETON MUDSTONES		
Mudstone, reddish brown with some greenish grey; halite veins near base	50.59	179.52
Halite, mostly clayey with thick partings of reddish brown and greenish grey mudstone	30.94	210.46
Mudstone, structureless and laminated, reddish brown	17.23	227.69
Mudstone, brecciated, mainly reddish brown	28.95	256.64
Mudstone, dominantly structureless, reddish brown with scattered, small, gypsum nodules	88.7	345.34
Mudstone, banded and blocky, reddish brown with some greenish grey	39.32	384.66
Mudstone, brecciated, grey and red (penecontemporaneous breccia at the level of the Rossall Salts)	27.73	412.39
Mudstone, mostly laminated, reddish brown with some greenish grey	27.74	440.13
HAMBLETON MUDSTONES		
Mudstone, partly interlaminated with siltstone, grey	15.24	455.37
Mudstone, brecciated, grey	5.79	461.16
Mudstone, reddish brown with thin band of sandstone	5.49	466.65
Mudstone, reddish brown, brecciated, with sandstone	4.27	470.92

SHERWOOD SANDSTONE GROUP
Sandstone, medium-grained, reddish brown with mudstone partings — 2.66   473.58

B5 Borehole [3628 3871] SD33NW/3

6.70 m above OD. See Figures 5 and 6. Drilled in 1954 for Imperial Chemical Industries. Cored from 48.77 to 291.39 m. Logged by D Thomas.

	Thickness m	Depth m
ALLUVIUM		
Clay, grey, soft	5.49	5.49
Gravel	2.13	7.62
GLACIAL DEPOSITS		
Boulder clay	5.79	13.41
Gravel	2.44	15.85
Boulder clay	11.58	27.43
THORNTON MUDSTONES		
Mudstone, red (rock-bitted)	21.34	48.77
Mudstone, red, brown and grey	15.85	64.62
SINGLETON MUDSTONES		
Mudstone, reddish brown with some greenish grey	46.63	111.25
Mudstone, structureless, brown; gypsum nodules	59.44	170.69
Mudstone, brown with some greenish grey	76.81	247.50
HAMBLETON MUDSTONES		
Mudstone, laminated in part, grey with some reddish brown near base	37.11	284.61
SHERWOOD SANDSTONE GROUP		
Sandstone, coarse-grained, grey, buff to grey and green	6.78	291.39

B6 Borehole [3487 4673] SD34NW/12

6.70 m above OD. Drilled in 1955–56 for Imperial Chemical Industries. Cored from 41.30 to 572.11 m. Logged by B J Taylor.

	Thickness m	Depth m
Unreliable driller's log to	41.30	41.30
COLLAPSE BRECCIA DUE TO COMPLETE SOLUTION OF SALT		
Mudstone, brecciated, red and grey	67.21	108.51
BRECKELLS MUDSTONES		
Mudstone, dominantly structureless, gypsum nodules common, some halite veins in lower part	143.86	252.37
COAT WALLS MUDSTONES		
Mudstone, laminated and structureless, dominantly reddish brown with some greenish grey	110.34	362.71

PREESALL SALT

Halite	5.49	368.20
Mudstone, grey, with halite crystals	0.30	368.50
Halite, slightly marly in part	42.68	411.18
Mudstone, red and grey, with anhydrite nodules	0.99	412.17
Halite, slightly marly	14.55	426.72
Mudstone, red and grey, with bands and crystals of halite	5.79	432.51
Halite, marly at intervals	76.20	508.71
Mudstone, brecciated, red and grey	0.61	509.32
Halite	29.87	539.19
Lost core (probably halite)	3.66	542.85
Marl	0.61	543.46
Halite, marly in part	2.13	545.59

THORNTON MUDSTONES (core not seen by BGS)

Mudstone, red with some grey; scattered beds of halite up to 1.2 m thick	12.19	557.78
Mudstone, red and grey	8.53	566.31
Mudstone, red, brecciated in part	5.80	572.11

B8 Borehole [3225 4529] SD34NW/18

4.90 m above OD. See Figures 5 and 6. Drilled in 1956 for Imperial Chemical Industries. Cored from 76.20 to 373.68 m. Logged by B J Taylor. Dip averages 4°.

	Thickness m	Depth m
MARINE ALLUVIUM		
Clay, sandy	7.32	7.32
Clay and gravel	0.91	8.23
Clay, sandy, gravel and shells	7.92	16.15
Gravel, sandy	10.98	27.13
BOULDER CLAY		
Clay and pebbles	7.31	34.44
THORNTON MUDSTONES		
Mudstone, mostly red (driller's log)	41.76	76.20
SINGLETON MUDSTONES (INCLUDES MYTHOP AND ROSSALL SALTS)		
Mudstone, red and grey	49.30	125.50
Mythop Salts		
Halite with much interbedded red mudstone	16.63	142.13
Mudstone, red and grey	18.50	160.63
Halite, marly	6.55	167.18
Mudstone, red; many halite veins	3.96	171.14
Halite	18.14	189.28
Mudstone, red, both structureless and banded	18.52	207.80
Mudstone, red, mostly structureless; anhydrite nodules	50.06	257.86
Mudstone, red, laminated with some structureless	29.34	287.20
Rossall Salts		
Halite, with dominant mudstone near top	11.50	298.70
Mudstone, red and grey	36.58	335.28
HAMBLETON MUDSTONES		
Mudstone, grey, laminated	17.68	352.96
Mudstone, grey, brecciated	8.23	361.19
Mudstone, grey	0.76	361.95
Mudstone, red with a few sandstone bands	7.11	369.06
SHERWOOD SANDSTONE GROUP		
Sandstone, reddish brown with a few mudstone bands	4.62	373.68

Churchtown Borehole [3256 4056] SD34SW/11

7.3 m above OD. See Figure 7. Drilled in 1969–70 for British Geological Survey. Cored from 32.94 to 151.48 m and stored at National Geosciences Data Centre, BGS, Keyworth. Logged by A A Wilson. Dip averages 8°.

	Thickness m	Depth m
GLACIAL DEPOSITS		
Boulder Clay	15.32	15.32
Sand	4.26	19.58
Boulder Clay	8.34	27.92
Sand	1.67	29.59
Boulder Clay	3.35	32.94
COLLAPSED BEDS AND BRECCIA DUE TO COMPLETE SOLUTION OF PREESALL SALT		
Mudstone, reddish brown with numerous veins and thick layers of gypsum	8.07	41.01
Mudstone, reddish brown with much micro-brecciation; gypsum veins	19.70	60.71
Mudstone, reddish brown	2.46	63.17
Mudstone, reddish brown, microbrecciated at several levels; gypsum veins and lenticles	6.25	69.42
THORNTON MUDSTONES		
Mudstone, often vividly banded in reddish brown and greenish grey, with siltstone bands; thin layers of gypsum and pods of anhydrite; *Euestheria* at 122.81 m	81.76	151.18

Coat Walls Borehole [3551 4654] SD34NE/30

7.5 m above OD. See Figure 13. Drilled in 1974 for British Geological Survey, Biostratigraphy Research Group. Cored from 20.00 to 285.93 m. Cores from 144 to 285.93 m stored at National Geosciences Data Centre, BGS, Keyworth. Logged by A A Wilson. Dip averages 17°.

	Thickness m	Depth m
GLACIAL DEPOSITS		
Clay, sandy, stony brown	4.80	4.80
Sand, scattered pebbles	0.90	5.70
Clay, sandy, stony, reddish brown	12.50	18.20
BRECKELLS MUDSTONES		
Mudstone, reddish brown, structureless with rare greenish grey blotches and bands; thin calcite veins at four restricted levels; gypsum veins absent	36.50	54.70
Mudstone, reddish brown, structureless with a few greenish grey bands; botryoidal gypsum nodules at many levels; gypsum veins usually present	39.70	94.40
Mudstone, reddish brown, structureless with rare laminated, greenish grey bands, becoming more abundant towards base; gypsum veins	57.90	152.30
COAT WALLS MUDSTONES		
Mudstone, reddish brown, dominantly structureless, alternating with units of greenish grey, largely laminated units, many of them greenish grey becoming progressively more common downwards; botryoidal gypsum nodules at a few levels but gypsum veins almost entirely absent.		

	Thickness m	Depth m
A few halite bands and crystal layers occur below 275.52 m	127.15	279.45
Mudstone, greenish grey with some reddish brown, brecciated in part; halite crystals near base	2.01	281.46

PREESALL HALITE

	Thickness m	Depth m
Halite with minor greenish grey mudstone wisps	4.47	285.93

E1 Borehole [3467 4746] SD34NW/4

9.50 m above OD. See Figure 11. Drilled in 1940 for Imperial Chemical Industries. Cored from 38.71 m. Logged by ICI.

	Thickness m	Depth m
GLACIAL DEPOSITS		
Boulder Clay	6.10	6.10
Sand	2.43	8.53
Boulder Clay	30.18	38.71
BRECKELLS MUDSTONES		
Marl, red and blue with gypsum	144.16	182.87
Rock-salt	0.02	182.89
Marl, red and blue with gypsum	9.13	192.02
COAT WALLS MUDSTONES		
Marl and shale, red and blue	99.06	291.08
Marl, red and grey with sporadic salt	26.37	317.45
PREESALL SALT		
No core	5.49	322.94
Rock-salt, marly in part (C₂ Bed?)	15.08	338.02
Marl, red	1.83	339.85
Rock-salt with marl (C₂ Bed?)	3.96	343.81
Marl, red	0.46	344.27
Rock-salt with some marl (C₁ Bed?)	7.16	351.43
Marl, red and grey	3.81	355.24
Rock-salt, marly in part (B Bed?)	9.15	364.39
Marl	1.21	365.60
Rock-salt, marly in part (B Bed?)	16.47	382.07
Marl, grey	1.37	383.44
Rock-salt (A Bed?)	14.78	398.22
THORNTON MUDSTONES		
Marl, red and grey	26.52	424.74
Marl, grey and rock-salt	0.15	424.89
Marl, red and grey	7.32	432.21
Marl, grey	9.29	441.50

E5 Borehole [3554 4419] SD34SE/2

11.60 m above OD. See Figures 7 and 11. Drilled in about 1940 for Imperial Chemical Industries. Cored from 64.16 to 272.47 m. Logged by ICI.

	Thickness m	Depth m
GLACIAL DEPOSITS		
Boulder clay with bands of sand and gravel	48.16	48.16
COAT WALLS MUDSTONE		
Mudstone, reddish brown with some greenish grey bands	50.90	99.06
PREESALL SALT		
Halite with several bands of mudstone and halite-mudstone rock	94.79	193.85

THORNTON MUDSTONE

	Thickness m	Depth m
Mudstone, reddish brown with some greenish grey bands	18.15	212.00
Halite, with intermingled mudstone	4.87	216.87
Mudstone, greenish grey and blue, alternating with some reddish brown	57.60	274.47

E6 Borehole [3631 4508] SD34NE/30

12 m above OD. Drilled in 1934 for Imperial Chemical Industries. Cored from 51.05–383.90 m. Logged by ICI.

	Thickness m	Depth m
MARINE ALLUVIUM		
Clay	1.07	1.07
Sand	0.45	1.52
GLACIAL DEPOSITS		
Boulder clay	5.03	6.55
Gravel	1.22	7.77
Boulder clay	4.57	12.34
Sand and gravel	0.61	12.95
Boulder clay	7.93	20.88
Sand	3.35	24.23
Boulder clay	18.29	42.52
THORNTON MUDSTONES?		
Marl, red, blue and grey	40.23	82.75
Marl, grey	15.55	98.30
SINGLETON MUDSTONES?		
Marl, red with some grey; gypsum throughout and slight traces of salt (Mythop Salts?); bedding horizontal at 183.95, almost vertical at 240.33 and 271.12	172.82	271.12
Marl, red and grey with gypsum; bedding almost vertical at 297.64, 60° at 318.97–354.94, almost vertical at 354.94–374.92	101.80	372.92
No core; sand, grey and salt (Rossall Salt?) recorded	10.98	383.90

Hackensall Hall Borehole [3498 4679] SD34NW/61

7.0 m above OD. See Figure 13. Drilled in 1974 for British Geological Survey, Biostratigraphy Research Group. Cored from 30.63 to 182.50 m. Logged by A A Wilson. Dip averages 8°.

	Thickness m	Depth m
GLACIAL DEPOSITS		
Clay, sandy, stony	5.00	5.00
Sand, brown, fine-grained, silty	7.00	12.00
Clay, dark brown, stony	4.50	16.50
Gravel, coarse	1.50	18.00
Clay, dark brown, stony	3.83	21.83
BRECKELLS MUDSTONES		
Marl, very hard, sandy with traces of gypsum (rock-bitted)	8.80	30.63
Mudstone, reddish brown, structureless; gypsum veins	4.82	35.45
Mudstone, reddish brown with some greenish grey patches and bands, brecciated, with gypsum porphyroblasts and irregular gypsum veins; three intervals with only very slight brecciation (collapse breccia on inferred, totally dissolved, multiple halite)	65.95	101.40

Mudstone, reddish brown, structureless with rare greenish grey bands from 130.10 to 182.50 m; botryoidal gypsum nodules at many levels — 81.10 — 182.50

Hambleton Borehole [3820 4217] SD34SE/5

9.8 m above OD. See Figures 5 and 6. Drilled in 1970 for British Geological Survey. Cored from 29.26 to 142.42 m and stored at National Geosciences Data Centre, BGS, Keyworth. Logged by A A Wilson. Dip averages 8°.

	Thickness m	Depth m
GLACIAL DEPOSITS		
Boulder Clay	29.26	29.26
SINGLETON MUDSTONES (INCLUDES BRECCIAS RELATED TO ROSSALL SALTS)		
Mudstone, reddish brown with a few greenish grey bands and a few thin beds of breccia; many gypsum vein	27.33	56.59
Mudstone, reddish brown with some greenish grey bands, brecciated; gypsum veins	11.10	67.69
Mudstone, reddish brown with a few greenish grey bands, well-banded in part, a few thin bands of breccia; gypsum veins	20.70	88.39
Mudstone, reddish brown and greenish grey; rare bands of gypsum nodules and veins of gypsum	14.20	102.59
HAMBLETON MUDSTONES		
Mudstone, grey with rare red bands, silty towards top; some salt pseudomorphs; scattered plant fragments; *Euestheria* and organic trail at 123.14 m	28.09	130.68
Siltstone and mudstone, grey with much brecciation	5.08	135.76
SHERWOOD SANDSTONE GROUP		
Sandstone, grey and green, fine- to medium-grained; slight hydrocarbon staining	6.66	142.42

Mythop Borehole [3647 3499] SD33SE/1

12.2 m above OD. See Figures 6 and 7. Drilled in 1970 for British Geological Survey. Cored from 34.44 to 248.11 m and stored at National Geosciences Data Centre, BGS, Keyworth. Logged by A A Wilson. Dip averages 5°.

	Thickness m	Depth m
GLACIAL DEPOSITS		
Boulder clay	34.14	34.14
COLLAPSED BEDS AND BRECCIA DUE TO COMPLETE SOLUTION OF PREESALL SALT		
Mudstone, reddish brown but greenish grey in places, brecciated at many levels; gypsum veins	19.71	53.85
THORNTON MUDSTONE		
Mudstone, reddish brown and greenish grey in cyclic alternations, in places interbedded with siltstone; scattered thin brecciated bands; gypsum nodules and gypsum veins at several levels	100.51	154.36

SINGLETON MUDSTONES WITH MYTHOP SALTS

	Thickness m	Depth m
Mudstone, reddish brown with some greenish grey bands, structureless and laminated, with several bands of halite up to 2.20 m in thickness; halite veins	63.85	218.21
Mudstone, reddish brown, chiefly laminated with some brecciated bands	29.90	248.11

P1 Borehole [3593 4781] SD34NE/126

6.10 m above OD. See Figure 11. Drilled in about 1962 for Imperial Chemical Industries. Cored from 28.96 m. Logged by ICI.

		Thickness m	Depth m
MARINE ALLUVIUM			
Clay, grey	about	1.80	1.80
Sand and gravel		5.20	7.00
GLACIAL DEPOSITS			
Boulder clay	about	22.00	29.00
BRECKELLS MUDSTONES			
Marl, red with some grey; breccia from 101.50–121.92		92.92	121.92
Marl, red with some grey; salt crystals at a few levels		4.27	126.19
COAT WALLS MUDSTONES			
Marl, red and grey; salt crystals near top and base; salt veins (average dip of strata 8°)		123.14	249.33
PREESALL SALT			
Rock-salt (C_3 Bed)		10.36	259.69
Marl with salt		0.15	259.84
Rock-salt (C_2 Bed)		17.53	277.37
Marl, red		0.30	277.67
Rock-salt (C_1 Bed)		16.84	294.51
Marl, red and grey with salt veins		2.97	297.48
Rock-salt (B Bed); marl at top and bottom		44.81	342.29
Rock-salt with much marl and salt veins		3.81	346.10
Rock-salt (A Bed)		26.67	372.77
THORNTON MUDSTONES			
Marl, red and grey		2.90	375.67
Rock-salt		3.19	378.86
Marl, red and grey		3.81	382.67
Marl, red and grey with some salt		6.86	389.53
Marl, red and grey		13.42	402.95

Preesall 101 Borehole [3561 4519] SD34NE/2

Drilled in 1954 for Imperial Chemical Industries. Cored from 176.78–335.28 m. Logged by F M Trotter.

	Thickness m	Depth m
GLACIAL DEPOSITS		
Marl, sandy	9.14	9.14
Sandstone (sic)	9.15	18.29
Gravel	3.05	21.34
Marl, sandy	3.04	24.38
Gravel	3.05	27.43
Marl and gravel	12.19	39.62
BRECKELL MUDSTONES		
Marl	28.35	67.97

Coat Walls Mudstones

	Thickness m	Depth m
Marl	108.81	176.78
Marl, red and grey; salt crystals and veins near base	13.11	189.89

Preesall Salt

Rock-salt, slightly marly (C$_3$ Bed)	6.25	196.14
Marl, red and grey with salt	0.61	196.75
Rock-salt, marly	0.61	197.36
Marl, grey	0.30	197.66
Rock-salt (C$_2$ Bed), marly to 201.78 m	17.68	215.34
Marl, red with salt	0.23	215.57
Rock-salt (C$_2$ Bed), marly to 225.25 m and below 231.34 m	16.23	231.80
Marl, red and grey with thin bands of marly rock-salt	3.65	235.45
Rock-salt (B$_1$ Bed), mostly marly	49.54	284.99
Marl, grey with salt	0.76	285.75
Rock-salt, marly	1.37	287.12
Marl, red and grey with salt	1.53	288.65
Rock-salt (A Bed), much marl above 295.05 m and below 311.51 m	28.34	316.99

Thornton Mudstones

Marl, grey and red, with veins of salt and gypsum	2.44	319.43
Rock-salt, marly	2.44	321.87
Marl, red and grey with salt veins and gypsum	1.68	323.55
Rock-salt, marly	1.52	325.07
Marl, red, and salt	0.46	325.53
Rock-salt, marly	0.45	325.98
Marl, red, with gypsum and salt	0.61	326.59
Marl, red and grey, brecciated	1.07	327.66
Rock-salt, marly	0.30	327.96
Marl, red and grey, with gypsum and traces of salt	7.32	335.28

Staynall Borehole [3562 4438] SD34SE/8

23.0 m above OD. Drilled in 1971 for British Geological Survey. Cored from 41.32 to 58.72 m and stored at National Geosciences Data Centre, BGS, Keyworth. Logged by A A Wilson. Dip averages 16°.

	Thickness m	Depth m
Glacial Deposits		
Boulder clay	36.72	36.72
Sand and boulders	2.88	39.00
Sand, boulders and clay	1.50	40.50
Coat Walls Mudstones		
Mudstone, reddish brown and structureless, alternating with reddish brown and greenish grey mudstone commonly interlaminated with bands of siltstone, some of them dolomitic; gypsum veins throughout	18.22	58.72

Thornton Cleveleys Borehole [3314 4409] SD34SW/15

6.0 m above OD. See Figures 6 and 7. Drilled in 1982 as part of the geothermal study programme of the Hydrogeology Unit, British Geological Survey. Cored from 42.00 to 310.00 m and stored at National Geosciences Data Centre, Keyworth. Logged by A A Wilson. Dip averages 11°.

	Thickness m	Depth m
Marine Alluvium		
Clay, brown, sticky, stoneless	6.0	6.0
Glacial Deposits		
Clay, brown, with pebbles	16.0	22.0
Clay, sandy, with very numerous pebbles	3.0	25.0
Clay, brown, with pebbles	11.0	36.0
Clay, sticky, yellowish brown	2.0	38.0
Clay, brown, with pebbles	4.0	42.0
Collapsed beds and breccia due to complete solution of Preesall Salt		
Breccia, dominantly reddish brown mudstone, with gypsum bands and porphyroblasts, and some less disturbed mudstone bands	45.85	87.85
Thornton Mudstones		
Mudstone, reddish, brown, dominantly structureless with some lamination in upper part	7.60	95.45
Mudstone, reddish brown alternating with greenish grey, interlaminated with siltstone; pseudomorphs after halite, desiccation cracks, current-ripple lamination	109.65	205.10
Halite, pale brown with some colour banding, mostly with a little greenish grey mudstone	4.90	209.00
Singleton Mudstones (with Mythop Salts)		
Mudstone, reddish brown, structureless, with several bands of halite up to 3.83 m in thickness	13.48	222.48
Mudstone, reddish brown, structureless, passing down into laminated beds with siltstone bands downwards	31.92	254.40
Mudstone, reddish brown, with some greenish grey, structureless, passing down into laminated beds with siltstone bands downwards; numerous beds of halite up to 3.40 m in thickness	55.60	310.00

Weeton Camp Borehole [3888 3603] SD33NE/9

24.50 m above OD. See Figures 4 and 5. Drilled in 1982 as part of the geothermal study programme of the Hydrogeology Unit, British Geological Survey. Cored from 29.00 to 300.11 m and stored at National Geosciences Data Centre, BGS, Keyworth. Logged by A A Wilson. Dip averages 6°.

	Thickness m	Depth m
Glacial Deposits		
Stony clay	7	7
Clays and gravels	8	15
Stony clay	14	29
Singleton Mudstones		
Mudstone, reddish brown with rare greenish grey, structureless with some poorly laminated	33.00	62.0
Mudstone, reddish brown, laminated and poorly laminated	23.25	85.25
Breccia with some undisturbed bands, reddish brown with greenish grey, a little siltstone (collapse breccia of Triassic age—lateral equivalent of the Rossall Salts)	24.45	109.70

Mudstone, reddish brown with some greenish
grey, dominantly laminated; siltstone bands,
current-ripple lamination, pseudomorphs
after halite ... 18.58 128.28

HAMBLETON MUDSTONES

Mudstone, greenish grey with rare reddish
brown layers, interlaminated with siltstone
and very fine-grained sandstone; very large
salt pseudomorphs often with calcite,
current-ripple lamination common ... 5.67 153.95

Breccia of mudstone and siltstone clasts in
matrix of very fine-grained sandstone ... 3.47 157.42

Mudstone with siltstone bands, greenish grey ... 1.48 158.90

SHERWOOD SANDSTONE GROUP

Sandstone, grey, fine-grained, cross-bedded ... 10.85 169.75

Mudstone, greenish grey, with sinuous
injections of sandstone, chalcopyrite laminae
in lower beds ... 3.21 172.96

Sandstone, grey, dominantly reddish brown
downwards, fine- to coarse-grained; cross-
bedding; mudstone pebbles in lower half ... 64.19 237.15

Sandstone, mostly reddish brown, fine-
grained, cross-bedding common ... 62.96 300.11

Winter Gardens Borehole [3090 3620] SD33NW/1

18.30 m above OD. Drilled in 1905 by C Timmins who probably
logged the borehole.

	Thickness m	*Depth* m
GLACIAL DEPOSITS		
Boulder clay	23.16	23.16
Sand and gravel	26.22	49.38
Loamy clay	10.67	60.05
Gravel, coarse	11.27	71.32
Clay, sandy	1.53	72.85
Clay, stiff (possibly Mercia Mudstone)	4.11	76.96
Gravel, sandy (possibly cavings)	0.31	77.27
SINGLETON MUDSTONES (MYTHOP SALTS)		
Rock-salt	0.61	77.88

75

INDEX